Laboratory Manual

ESSENTIALS OF

Anatomy
& Physiology

Eighth Edition

Kevin T. Patton
St. Charles Community College

Mc Graw Hill

Connect
Learn
Succeed™

LABORATORY MANUAL SEELEY'S ESSENTIALS OF ANATOMY AND PHYSIOLOGY,
EIGHTH EDITION

Published by McGraw-Hill, a business unit of The McGraw-Hill Companies, Inc., 1221 Avenue of the Americas, New York, NY 10020. Copyright © 2013 by The McGraw-Hill Companies, Inc. All rights reserved. Printed in the United States of America. Previous editions © 2010, 2007, and 2005. No part of this publication may be reproduced or distributed in any form or by any means, or stored in a database or retrieval system, without the prior written consent of The McGraw-Hill Companies, Inc., including, but not limited to, in any network or other electronic storage or transmission, or broadcast for distance learning.

Some ancillaries, including electronic and print components, may not be available to customers outside the United States.

♻ This book is printed on recycled, acid-free paper containing 10% postconsumer waste.

3 4 5 6 7 8 9 0 QTN/QTN 1 0 9 8 7

ISBN 978–0–07–739131–7
MHID 0–07–739131–4

Vice President, Editor-in-Chief: *Marty Lange*
Vice President, EDP: *Kimberly Meriwether David*
Senior Director of Development: *Kristine Tibbetts*
Publisher: *Michael S. Hackett*
Executive Editor: *Colin H. Wheatley*
Director of Digital Content Development: *Barbekka Hurtt, Ph.D.*
Developmental Editor: *Mandy C. Clark*
Marketing Manager: *Chris Loewenberg*
Project Manager: *Mary Jane Lampe*
Senior Buyer: *Laura Fuller*
Designer: *Tara McDermott*
Cover/Interior Designer: *Elise Lansdon*
Cover Image: © *Tom Dahlin (thermographic image);* © *Ty Allison/Getty Images (triathlete)*
Senior Photo Research Coordinator: *John C. Leland*
Photo Research: *Jerry Marshall/pictureresearching.com*
Compositor: *Aptara®, Inc.*
Typeface: *10/12 Times Roman*
Printer: *Quad/Graphics*

All credits appearing on page or at the end of the book are considered to be an extension of the copyright page.

Some of the laboratory experiments included in this text may be hazardous if materials are handled improperly or if procedures are conducted incorrectly. Safety precautions are necessary when you are working with chemicals, glass test tubes, hot water baths, sharp instruments, and the like, or for any procedures that generally require caution. Your school may have set regulations regarding safety procedures that your instructor will explain to you. Should you have any problems with materials or procedures, please ask your instructor for help.

Contents

Preface

Anatomy and physiology laboratory courses provide the essential hands-on learning opportunities required for a thorough understanding of the human body. This lab manual contains a series of 43 exercises, which provide several guided explorations of human structure and function. The following activities are included:

o **Labeling exercises** provide opportunities to identify important structures learned in the laboratory and lecture portions of the course. Once completed, they provide guidance to laboratory examinations of models and specimens. Students are encouraged to *write out* the labels, to help them learn the terms more easily.

o **Coloring exercises** are noted as one of the most popular and most effective ways for many learners to grasp the essential spatial relationships of anatomical structures. This lab manual contains an accurate and comprehensive collection of human anatomy and physiology coloring plates.

o **Dissection of anatomical models** and examination of charts are an integral part of any beginning anatomy and physiology laboratory experience. This manual's instructions give valuable guidance for the effective use of models and charts.

o **Dissection of fresh and preserved specimens** of tissues, organs, and whole organisms enhances each student's appreciation of anatomical and functional relationships. Examination of tissues and organs is suggested in specific exercises throughout this manual. The last unit offers a complete dissection of the laboratory rat (which can be used at any point in the course), as well as a guide to a live or taped demonstration with a prosected human cadaver.

o **Physiological experiments** emphasizing a variety of functional processes of the human body offer students immediate and dramatic examples of physiological concepts. When possible, these activities center around examination of the student's *own* physiological processes.

o **Optional computerized experiments** allow students to use the latest and most accurate methods for observing and measuring important physiological phenomena.

o **Advanced scientific inquiry** projects encourage students to use computer simulations of real lab experiments. The computer simulations used for these projects are from McGraw-Hill's PHYSIOLOGY INTERACTIVE LAB SIMULATIONS (Ph.I.L.S. 4.0). Contact your campus bookstore or go online to www.mhhe.com.

o **A collaborative, group approach** to laboratory activities is emphasized throughout this manual to foster interactive learning processes and critical thinking.

o **Content and concept review** questions and fill-in tables, in lab reports and throughout the text of various exercises, encourage students to reinforce and apply their knowledge of human structure and function.

o **Modern anatomical imaging** techniques, such as computed tomography (CT), magnetic resonance (MR) imaging, and ultrasonography, are introduced where appropriate. In each special presentation of imaging technology, students are challenged to interpret actual images of the human body.

o **Practical applications** to exercise and athletics, clinical situations, and everyday experiences increase student motivation and place important concepts in a useful context. Each practical example includes application questions that encourage students to think about how concepts apply to the situation described.

This lab manual also offers other features that enhance learning and ensure a safe and effective laboratory experience:

o **Special reference supplements** are included where appropriate to help students in their examination of laboratory specimens. The Histology Minireference offers numerous full-color, labeled examples of commonly seen histology specimens. The "Anatomical Atlas of the Rat" in Exercise 41 includes a complete series of labeled anatomical drawings of the laboratory rat.

o **Learning objectives** presented at the beginning of each exercise offer a framework for learning.

o **A complete list of materials** for each exercise give the students and instructor a handy reference for efficient setup of laboratory activities.

- Boxed hints provide students with special tips on handling specimens, using equipment, and otherwise managing their laboratory activities.

- Safety tips are highlighted in special boxes to remind students of potential hazards, such as fire, chemical spills, cuts, and biological contamination.

- Numerous illustrations of proper procedures complement the text's complete description of laboratory activities.

The design of this laboratory package not only makes the laboratory course fun and effective for the student but also provides essential support for the instructor and lab preparation technician. Here are some examples of elements designed to aid instruction and preparation:

- A comprehensive Instructor's Manual and preparation guide is available for instructors at www.mhhe.com/labcentral. The guide contains a complete set of hints and special notes and instructions for each laboratory exercise. The guide provides a list of materials broken down by exercise and a comprehensive list of all materials suggested for the course. Substitutions and special sources are given where appropriate. Lists of solution-preparation guidelines and other aids are found throughout the guide. Reproducible handouts to supplement certain exercises are also provided in the guide.

- Modular organization of laboratory exercises allows their use in virtually any order required by the needs of individual courses. Comprehensive cross-references in the instruction and preparation guide alert instructors about other lab exercises that may involve similar material. For example, the exercise on hormones (Lab Exercise 24) may be more appropriately done near the reproductive-system exercises in some courses. The major dissection unit is isolated at the end of the manual for easy reference if dissection specimens are to be used throughout the course or at any particular point in the course.

- Complete instructions for lab activities are given to the student, freeing the instructors to interact with laboratory students on an individual basis rather than spending a great deal of time introducing the lab procedure to the whole class.

- Easy-to-evaluate lab report formats allow instructors to check on student work or assign grades in an efficient manner. The Instruction and Preparation Guide provides correct answers to objective questions in each lab report.

Acknowledgments

Production of this lab manual was a team effort in many ways. Illustrator Eileen Draper contributed many of the coloring plates and other figures. Tracy Powers, Amy White, Victoria Ferrario, Mary Atchison, and Stacey Thater, my lab gurus, provided invaluable advice over the years for improving and refining lab activities. Authors Cinnamon VanPutte, Jennifer Regan, and Andy Russo provided an excellent textbook on which to base this lab manual, as well as helpful encouragement and advice. Thanks, too, for the support of Developmental Editor Mandy Clark and Project Manager Mary Jane Lampe at McGraw-Hill Publishers.

My sincerest thanks to everyone involved.

Kevin Patton

Additional Resources

Anatomy & Physiology I REVEALED® Student Tutorial

Anatomy & Physiology I REVEALED® is a unique multimedia study aid designed to help students learn and review human anatomy using digital cadaver specimens. Dissections, animations, imaging, and self-tests all work together as an exceptional tool for the study of structure and function.

Anatomy & Physiology I REVEALED® includes:

- Integumentary System

- Skeletal and Muscular Systems

- Nervous System

- Cardiovascular, Respiratory, and Lymphatic Systems

- Digestive, Urinary, Reproductive, and Endocrine Systems

- Expanded physiology content

- Histology material

Visit www.mhhe.com/APREVEALED.com for more information.

Textbook Website

The website that accompanies *Seeley's Essentials of Anatomy and Physiology* at www.mhhe.com/seeleyess8 allows instructors to browse, select, and export files containing artwork from the text in multiple formats to create

McGraw-Hill ConnectPlus® Anatomy & Physiology interactive learning platform provides a customizable, assignable eBook, autograded assessments, an adaptive diagnostic tool, lecture capture, access to instructor resources, and powerful reporting—all in an easy-to-use interface.

Learn more at www.mcgrawhillconnect.com.

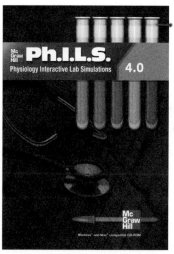

customized classroom presentations, visually based tests and quizzes, dynamic course website content, or printed support materials. The digital assets on the website are grouped by chapter within easy-to-use folders.

o **Art** Full-color digital files of all illustrations in the book and unlabeled versions of the same artwork can be readily incorporated into lecture presentations, exams, or custom-made classroom materials. In addition, all files are pre-inserted into blank PowerPoint slides for easy lecture presentations.

o **Photos** Digital files of instructionally significant photographs from the text can be reproduced for multiple classroom uses.

o **Tables** Every table that appears in the text is available to instructors in electronic form.

o **Animations** Numerous full-color animations illustrating physiological processes are provided. Harness the visual impact of processes in motion by importing these files into classroom presentations or online course materials.

Students will benefit from **practice quizzing, animation quizzing,** and other **study tools,** all correlated by chapter. Help with difficult concepts is only a click away!

NEW! Ph.I.L.S. 4.0 has been updated! Users have requested, and we are providing, *five new exercises* (Respiratory Quotient, Weight and Contraction, Insulin and Glucose Tolerance, Blood Typing, and Anti-Diuretic Hormone). Ph.I.L.S. 4.0 is the perfect way to reinforce key physiology concepts with powerful lab experiments. Created by Dr. Phil Stephens at Villanova University, this program offers *42 laboratory simulations* that may be used to supplement or substitute for wet labs. All 42 labs are self-contained experiments—no lengthy instruction manual required. Users can adjust variables, view outcomes, make predictions, draw conclusions, and print lab reports. This easy-to-use software offers the flexibility to change the parameters of the lab experiment. There are no limits!

Welcome to Anatomy and Physiology Laboratory

Anatomy and physiology laboratory challenges you to learn a great deal about human structure and function in a rather informal, practical atmosphere. Despite its informality as a learning situation, laboratory work does require some appreciation of the scientific method in general and laboratory policy and procedure in particular. Read this introductory section carefully. It will introduce you to laboratory science and will give you tips on how to complete this course with great success.

The Scientific Method

The **scientific method** is merely an approach to discovery. From its early days as a discipline, science has relied on this very simple, logical method for gaining an understanding of the universe. The basics of the scientific method can be summarized as the following set of steps that are followed in scientific discovery.

Hypothesis.   First, one makes a tentative explanation, called a **hypothesis,** about some aspect of nature. A hypothesis is a reasonable guess based on previous informal observations or on previously tested explanations.

Initial experimentation.   After a hypothesis has been proposed, it must be tested. The testing of a hypothesis is called **experimentation.** Scientific experiments are designed to be as simple as possible, to avoid the possibility of errors. Often, **experimental controls** are used to ensure that the test situation itself is not affecting the results. For example, if a new cancer drug is being tested, half the test subjects will get the drug and half the subjects will be given a harmless substitute. The group getting the drug is called the *test group,* and the group getting the fake is called the *control group.* If both groups improve, or if only the control group improves, the drug's effectiveness hasn't been proven. If the test group improves, but the control group doesn't, the hypothesis that the drug works is tentatively accepted as true. Experimentation requires accurate measurement and recording of data.

Interpretation and conclusion.   After an experiment, or a series of experiments, the researcher analyzes all of the experimental data. If the results support the original hypothesis, it is tentatively accepted as true, and the researcher moves on to the next step. If the data do not support the hypothesis, the researcher tentatively rejects the hypothesis. If an experimental error is suspected, the hypothesis might be retested instead of rejected. Knowing which hypotheses are untrue is almost as valuable as knowing which are true. Every rejected hypothesis brings the scientific community a little closer to the truth.

Replication.   This step in the scientific method is the one least appreciated by nonscientists. After a hypothesis is tested and accepted, it is retested again and again to make sure that it is true. Because researchers often make minor mistakes in the design or execution of experiments, it is important that other scientists verify the original work. Usually, initial research experiments and their results are published in scientific journals so that others in the same field of research can benefit from them and verify them. If experimental results cannot be replicated (recreated) by other scientists, the hypothesis is not widely accepted. If a hypothesis withstands this rigorous retesting, the level of confidence in the hypothesis increases. A hypothesis that has gained a high level of confidence is called a **theory** or **law.**

The "facts" presented in this course are merely the latest hypotheses of how the body is built and how it functions. As methods of imaging the body and measuring functional processes improve, we find new data that cause us to replace old hypotheses with newer ones.

Measurement and Data Collection

Scientific experimentation is valid only when data are accurately measured and accurately recorded. In this lab course, you will be invited from time to time to execute various experiments. Your results will be meaningful only if you are careful to measure and record your results properly.

In this manual, as in nearly all scientific works, only the **metric system** of measurement is used. The metric system is useful for two important reasons: it is commonly used throughout the world, and metric units are easily converted because they are all based on units of 10.

In this course, you will use metric units to measure time, temperature, length, volume, mass, and pressure. Each of these measurable characteristics has a basic metric unit that can be increased or decreased by factors of 10 as needed. For example, the basic unit of measuring length is the **meter.** The meter is a useful unit in measuring the height of an elephant, 2 to 3 meters, but not in measuring the distance to the moon—for this, units a thousand times larger, **kilometers,** are used instead. Likewise, the diameter of a bacterial cell is measured in **micrometers,** which are a million times smaller than a meter. Different-size units can be converted back and forth by multiplying or dividing by factors of 10.

Each of the tables in this section gives the basic units and common alternative units for time, temperature, length, volume, mass, and pressure.

Time

Basic unit: second (sec.)

0.000001 second = microsecond (μsec.)

0.001 second = millisecond (msec.)

60 seconds = minute (min.)

3,600 seconds = hour (hr.)

Temperature

Basic unit: degree Celsius (°C)

No alternate units are commonly used.

Length

Basic unit: meter (m)

0.000001 meter = micrometer (μm)

0.001 meter = millimeter (mm)

0.01 meter = centimeter (cm)

1,000 meters = kilometer (km)

Volume

Basic unit: liter (l or L)

0.001 liter = milliliter (mL)*

0.01 liter = centiliter (cL)

0.1 liter = deciliter (dL)

*milliliters = cubic centimeters (cc)

Mass

Basic unit: gram (g)

0.001 gram = milligram (mg)

0.01 gram = centigram (cg)

1,000 grams = kilogram (kg)

Pressure

Basic unit: millimeters of mercury (mm Hg)

No alternate units are commonly used.

Accurate measurement means the correct use of units of measurement, but it also means the accurate use of measuring devices. If you are not familiar with reading the markings on metric rulers, balances, thermometers, and other common measuring devices, ask your instructor to demonstrate.

How to Dissect

The term *anatomy* literally means "to cut apart," so it is no wonder that dissections are commonly done in anatomy lab courses. Dissection activities recommended in this manual are not proposed without recognition of humane concerns. The purpose of these dissections is to instruct in a way that no other method can duplicate. The specimens called for are from animals raised specifically to be euthanized and used as resources. For health professionals, dissection of anatomical preparations is a necessary prerequisite to working with living bodies.

The goal of any dissection exercise is the exploration of anatomical relationships. Proper dissection requires patience and skill. Some students slice and hack away at their specimens until they have a tray of ground meat. Others hardly touch their specimen. Good technique is somewhere in the middle of these two extremes. Organs should be separated from one another only enough to see surrounding structures. Rarely should structures be cut or removed. The instructions given in this manual state when, where, and how cuts should be made.

Each dissection activity in this manual offers safety advice concerning proper handling of the dissection specimens. Protective gloves, lab coat, and eyewear are recommended. Always be careful when using dissection tools. Severe injuries can result from their careless use.

Figure A shows some commonly used dissection instruments. A brief discussion of each is in order here:

☐ **1** **Scalpel or knife**—The scalpel is probably the most overused instrument in an anatomy lab course. This tool should be used *seldomly*, only when you want to cut all the way through a specimen. A pathology knife, or butcher knife, is a much more useful tool.

☐ **2** **Scissors**—Scissors are probably the most underused dissection tool. Whenever you are tempted to use a scalpel for cutting, try the scissors first. Often your results will be much better and you will not have damaged important underlying parts.

☐ **3** **Probes**—You may want to have several types of probes. Both dull and sharp probes (**dissecting needles**) are useful in separating tissues, exploring cavities, tracing blood vessels, and pointing to structures.

☐ **4** **Forceps**—Forceps are one of the handiest dissection tools. They can be used to grasp small objects, to separate structures, to point to structures, to explore cavities, and to pull on structures.

☐ **5** **Ruler**—Your metric ruler should be marked in both centimeters and millimeters. It is useful in measuring organs and in many nondissection lab activities.

☐ **6** **Dissection pins**—These are standard, heavy-duty straight pins. Dissection pins are useful in pinning membranes and other structures to a dissection board to keep them temporarily out of the way.

☐ **7** **Dissection tray**—There are many varieties and shapes of dissection trays, and your instructor will recommend the best for your situation. Dissection trays help organize the dissection activity by keeping everything together, and they protect the lab table surface. Some types can also be used for storing your specimen for later study.

| Scalpel | Knife | Scissors | Dissecting needle | Dull probe | Forceps |

Figure A   Commonly used dissection tools.

Using This Manual

The format of each lab exercise is self-explanatory. After some brief introductory remarks, each exercise begins with a section entitled "Before You Begin." This section recommends that, before starting any lab activity, you

o   Read the appropriate chapter(s) in your textbook.

o   Set your learning goals so that you know what important concepts you should be trying to learn.

o   Prepare your materials so that your lab activity will run smoothly.

o   Read the entire activity before starting. In this way, you won't be surprised by (and therefore unprepared for) any step in the procedure.

Each exercise contains one or more "activities." Your instructor may suggest that you do all the activities in an exercise or only one. Each activity includes some or all of these helpful features:

o   Large, bold step numbers help you keep your place as your eyes move back and forth between the manual and your lab setup. Each step number and substep is preceded by a check-box (or hollow bullet), which you can use to check off each step and substep as it is completed.

o   Boxed hints, safety tips, and landmark characteristics (for identifying specimens) highlight useful information for completing the activity safely and successfully.

o   Labeling exercises encourage you to apply your knowledge of human anatomy in a practical test. Long label lines allow you room to write the name of the structure directly on the illustration. When each figure is completely labeled, the terms should be transferred to the lab report so that your instructor can check your work.

o   Coloring exercises have been cited by educators and students as being a very effective method of learning anatomy. By using multiple senses to trace and highlight shapes and relationships, you reinforce your knowledge of the human form. Each coloring plate is like a paint-by-number activity. Each label has a small number corresponding to a number on the illustration itself. Use a colored pen or pencil (not a crayon) to write the term over the outline letters of the label. This will reinforce your familiarity with the term and its correct spelling and makes the label a "color code" so you can later find the structure by its color. After filling in the label, use the same color to shade in the matching structure in the illustration. Use contrasting colors so that parts that are

next to one another can be easily distinguished. Do not attempt to color everything in realistic colors; otherwise, many parts will look the same. If you use pens, make sure they won't bleed through the paper. If you use pencils, you may want to insert a piece of plain paper into your manual over each completed coloring plate so that the colors don't rub off on the facing page. The materials list in the first exercise reminds you to bring colored pens or pencils to your work area, but it is assumed that you will have these for all the lab exercises.

Each exercise ends with a lab report. The report includes handy tables and sketching areas for you to record your laboratory observations and results. Where appropriate, the lab report also presents objective and subjective questions for you to answer. Some of these questions drill you on your knowledge of important terms, whereas others ask you to use your basic knowledge to interpret, organize, or otherwise process basic facts to ensure that you understand conceptual relationships. A number of highlighted practical application boxes in the exercise itself often ask you to apply your basic knowledge in this way. These questions, along with practice in making scientific observations, help you develop your scientific reasoning skills.

Safety First!

As already mentioned, numerous boxed safety tips—entitled "Safety First!"—appear throughout this manual. Their large letters and urgent tone are meant to call your attention to potential hazards in the laboratory. Before beginning the course, it is essential that you learn some basic laboratory safety policies (fig. B):

o   Always use care and common sense in the laboratory.

o   Check the labels of all chemicals for safety warnings before using them. If the container doesn't have a safety label, consult your instructor.

o   Wear protective gear if you are handling dangerous materials.

o   Plan ahead about what to do if a spill occurs. If you spill a dangerous substance on your body, you may have to remove your clothing. If your lab has a safety shower, stand under it and pull the ring.

o   Avoid using materials to which you have a known sensitivity.

o   Locate first aid equipment in the laboratory and familiarize yourself with its use.

o   Locate the nearest emergency medical help, and identify the easiest and fastest way to access it (for example, phoning an ambulance).

o Locate the fire safety equipment in your lab and familiarize yourself with its use. Locate the nearest fire alarm box, and identify the recommended primary and alternate fire exit routes.

o Confirm all directions, procedures, and other information with your instructor prior to each activity. Mark changes or additions in this manual, crossing out any information that does not apply to you.

o **Always** follow the directions provided with equipment and supplies, even if they are different from those given in this manual. Injuries often result from the misuse of equipment.

o Remember that absolutely no eating, drinking, or smoking is allowed in the laboratory.

o To avoid injury and/or contamination, properly dispose of, or clean and store, all lab equipment and supplies before leaving the laboratory. Wash your hands thoroughly on entering and leaving a biological laboratory.

o Always supervise experiments or demonstrations in progress. Never leave a laboratory experiment unattended.

Chemical Safety

Safety shower

Pull

Eyewash
(hold eyelids open)

Safety goggles

Protective gloves

Fire Safety

1. Pull pin

2. Aim

3. Squeeze handles together

Fire extinguisher

Fire blanket
(wrap yourself as
you pull it out)

Figure B Commonly used chemical and fire safety equipment.

Important Safety Information

Use the following spaces to write in important safety information.

Location of first aid box:

Emergency medical help:

Fire extinguisher location:

Fire extinguisher type:

Fire alarm box location:

Primary escape route:

Alternate escape route:

Weather safety shelter:

Important notes:

How to Study for Lab Quizzes

Tips for effective studying could fill several volumes. Instead of an entire work on studying, this section presents a few tips that many students have found to be particularly helpful in the anatomy and physiology laboratory course. As with any list of study tips, you will find some useful to you and others not so useful.

o   Identify the type of quizzes you will encounter. Written quizzes may involve multiple-choice, matching, or other objective questions. Written quizzes may also have subjective short-answer or essay questions in which you are to interpret or restate concepts. Figures may be offered for labeling. Practical tests may involve stations at which specimens have been placed. You may be asked to perform a procedure or answer a question about the specimen while at each station. You may have an individualized practical quiz in which your instructor will ask you questions or watch you perform a procedure individually. Ask your instructor for some sample questions if you don't understand how an upcoming quiz is constructed. Knowing the manner with which you will be evaluated is important to designing your study strategy.

o   Organize a study group (which may be the same as your lab group) to meet several times before a quiz to help each other with the material. Quizzing each other is a useful technique.

o   Spend many brief sessions reviewing specimens and other lab work, rather than one or two long sessions just before the quiz. Ask your instructor if there are open lab times or a learning center at which you can review models or specimens.

o   Study actively. Don't just sit and look at your study materials but do some hands-on activities with them. Make flash cards to quiz yourself. Set up your own practical test; then take it. Sketch specimens and models again. Draw a "concept map" in which you draw boxes or pictures connected in a logical way. This will help you organize concepts in your mind.

o   Use all your resources. Campus libraries and learning centers often offer the services of specialists who can help you sharpen your study skills. They can show you examples of concept maps and flash cards. Your instructor has successfully completed laboratory courses and has helped numerous previous students in this course. Ask your instructor for study tips. Use the textbook and study aids in the textbook to help you learn the lab material.

Histology Minireference

Plate 1
Mitosis (whitefish).
(*a*) Prophase, (*b*) metaphase, (*c*) anaphase, (*d*) telophase, (*e*) interphase.

(a)

Chromosomes
condensed

Spindle fibers

(b)

Chromosomes at
cell's equator

Spindle fibers

(c)

Centromeres
separating

(d)

Cells divide
into two cells

Dispersed chromosomes
form chromatin

Nuclear
envelope

(e)

Plate 2
(*a*) Simple squamous epithelium (surface view), (*b*) simple cuboidal epithelium (human),
(*c*) simple columnar epithelium (human uterus).

(a)

Squamous
epithelial cell

Nucleus

Nucleus

Basement membrane

Lumen
(open space)

Free surface

Simple cuboidal
epithelial cell

(b)

Free surface

Epithelial cell

Nucleus

Nucleolus

Basement membrane

(c)

Plate 3
(*a*) Stratified squamous epithelium (human vagina), (*b*) stratified cuboidal epithelium (ovarian follicle),
(*c*) stratified columnar epithelium (human).

Basement   Nuclei   Stratified squamous   Free surface
membrane   epithelial cell

(a)

Stratified cuboidal epithelial cells   Ovum

(b)

Columnar epithelial
cells at free surface   Free surface

Basement membrane   Epithelial
cells

(c)

Plate 4
(*a*) Transitional epithelium (human ureter), (*b*) pseudostratified epithelium,
(*c*) glandular epithelium (exocrine).

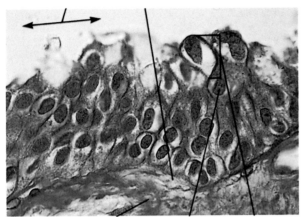

Free surface   Basement membrane

Connective tissue   Nucleus   Transitional epithelial cell

(a)

Free surface   Cilia at free surface   Basement membrane   Nucleus

(b)

Basement membrane   Free surface   Goblet cell

(c)

Plate 5

(*a*) Areolar (loose) connective tissue, (*b*) dense fibrous (irregular collagenous) connective tissue, (*c*) dense regular (elastic) connective tissue, (*d*) adipose tissue.

(a)

Bundle of collagen fibers

Nucleus of fibroblast

Elastic fibers

Nucleus of fibroblast

(b)

Irregularly arranged collagen fibers

(c)

Nucleus   Elastic fibers

(d)

Storage area for fat

Cell membrane

Nucleus of adipose cell

Plate 6
(*a*) Reticular connective tissue, (*b*) hyaline cartilage, (*c*) fibrocartilage,
(*d*) elastic cartilage.

Blood cell   Reticular fibers

Cell nuclei

(a)

Chondrocyte   Lacuna   Matrix

(b)

Chondrocyte   Lacuna   Collagen fibers
in matrix

(c)

Elastic fibers   Chondrocyte   Lacuna
in matrix

(d)

Plate 7
(*a*) Skin (human), (*b*) skin (human palm), (*c*) scalp (human),
(*d*) sweat gland.

(a)

Epidermis

Keratinized stratified
squamous epithelial cells
(stratum corneum)

Dermis

Germinal layers of
epithelial cells

Duct of sweat gland

Stratum corneum

Dermis

Papillae of dermis

Epidermis

(b)

Epidermis

Dermis

Hypodermis

Hair
follicle

Sebaceous
gland

(c)

Epidermis

Dermis

Duct of
sweat gland

Sweat gland
(coiled tube)

(d)

Plate 8
(*a*) Cancellous bone, (*b*) compact bone, (*c*) periosteum.

Matrix   Bone marrow   Nucleus

(a)

Canaliculi   Lacuna (contains osteocyte)

(b)

(c)   Periosteum   Bone

Plate 9

(*a*) Intramembranous bone formation (human), (*b*) endochondral bone formation (monkey), (*c*) endosteum.

(a)

Osteocytes Osteoblasts

(b)

Chondrocytes enlarging

Bone trabeculae Osteoblasts Calcified cartilage Chondrocytes in lacuna

Medullary cavity Epiphyseal plate Hyaline cartilage

Osteoblast Endosteum Marrow

Osteocyte Matrix

(c)

Plate 10
(*a*) Haversian system, (*b*) trabecula, (*c*) sharpey's fibers.

Lacuna (contains osteocyte)   Canaliculi

Osteon or Haversian system   Haversian or   Lamellae
                              central canal

(a)

Osteoblast   Trabecula   Osteocyte
                                       within lacuna

(b)

Perforating or Sharpey's fiber

Osteocyte   Bone matrix   Periosteum

(c)

Plate 11
(*a*) Skeletal muscle, (*b*) skeletal muscle, longitudinal and cross section, (*c*) cardiac muscle,
(*d*) visceral (smooth) muscle.

Striations

Skeletal muscle fiber (longitudinal section)

Skeletal muscle fiber   Nucleus   Nucleolus

(a)

Skeletal muscle fiber (cross section)   Nucleus

(b)

Nucleus

Nuclei of
smooth muscle cells

Striations   Intercalated disk

(c)

(d)

Plate 12
(*a*) Pituitary (anterior), (*b*) pituitary (posterior), (*c*) thyroid,
(*d*) adrenal.

Pars distalis   Capsule

(a)

Pars nervosa

(b)

Thyroid
follicle

Follicular
cells

Parafollicular
cells

Colloid in
follicle

Adrenal medulla

Adrenal cortex   Medullary veins

(c)

(d)

Plate 13
(*a*) Pancreas, (*b*) Graafian follicle (cat), (*c*) seminiferous tubule.

Endocrine portion of pancreas   Exocrine portion of pancreas
(islet of Langerhans)

(a)

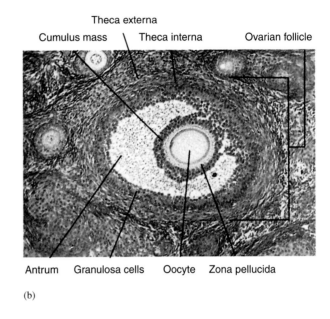

Theca externa

Cumulus mass   Theca interna   Ovarian follicle

Antrum   Granulosa cells   Oocyte   Zona pellucida

(b)

Basement membrane

Seminiferous tubule

(c)

Plate 14
(*a*) Lymph nodule, (*b*) lymph node, (*c*) spleen, (*d*) thymus.

Intestinal epithelium

Lymph nodule   Diffuse lymphatic tissue

(a)

Lymphocytes in lymph node   Capsule

Trabeculae

(b)

Red pulp   Blood vessels   White pulp

(c)

Lobule

Medulla

Cortex

Thymic corpuscle

(d)

Plate 15
(*a*) Blood smear, (*b*) blood (human), (*c*) lymphocyte,
(*d*) phagocytic white blood cell.

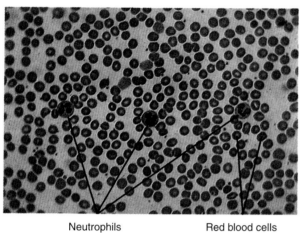

Neutrophils   Red blood cells

(a)

Red blood cell   Neutrophil   Platelet

(b)

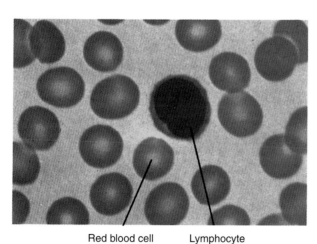

Red blood cell   Lymphocyte

(c)

Red blood cell   Neutrophil

(d)

Plate 16
(*a*) Unipolar neuron, (*b*) multipolar neuron, (*c*) neuroglia,
(*d*) spinal cord cross section (cat).

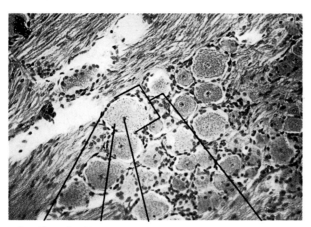

Satellite cells Nucleus Nucleolus   Nerve cell body
(a)

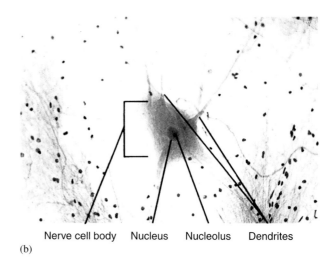

Nerve cell body   Nucleus   Nucleolus   Dendrites
(b)

Neuroglial cells (neuroglia)

Nerve cell body   Nucleus Nucleolus   Dendrites
(c)

Peripheral white portion   Posterior horn   Dorsal root

Anterior median fissure   Central canal   Anterior horn
(d)

LAB EXERCISE 1

The Microscope

In the seventeenth century, amateur Dutch scientist Anton van Leeuwenhoek used one of the first microscopes to discover a whole new world of living organisms. Using a single lens, or **simple microscope,** he observed tiny organisms in pond water and other substances. Robert Hooke, an English scientist, discovered that larger organisms had small microscopic subunits that he called "cells." Ever since this early era of discovery, biological microscopy has been essential in the study of living organisms.

The microscopes used in this course are **compound microscopes,** made of a set of lenses. They are more powerful, and more complex, than those used by van Leeuwenhoek and Hooke. This exercise will introduce you to the use and care of the standard compound microscope.

Before You Begin

☐ Read the appropriate chapter in your textbook.

☐ Set your learning goals. When you finish this exercise, you should be able to:
  o identify each major part of a compound light microscope and describe its function
  o determine total magnification at different settings
  o use a microscope to observe prepared specimens
  o prepare a wet-mount slide
  o stain microscopic specimens
  o accurately record microscopic observations

☐ Prepare your materials:
  o compound light microscope
  o prepared microslides (microscope slides):
        newsprint e
        three colored threads

  o clean slides
  o coverslips
  o paper wipes
  o methylene blue stain
  o flat toothpicks
  o disposable gloves
  o protective eyewear

☐ **Carefully** read the directions and safety tips for this exercise before starting any procedure.

A. Parts of the Microscope

☐ **1** Obtain a compound light microscope (fig. 1.1) and identify each of the structures described below.

Hint . . . Because not all microscope models are alike, some of the structures described below may be different on your microscope.

☐ **2** The set of lenses closest to your eye is the **ocular,** or eyepiece. The ocular magnifies an image by the factor indicated on the ocular's barrel, usually 10×. If the factor is 10×, the image is magnified 10 times. If the factor is 5×, the image is magnified 5 times. Many *binocular* microscopes (those with two oculars) have an ocular adjustment feature on one of the oculars. This feature allows you to adjust the focus of one ocular to match what you see in the other ocular (when focused). Many people have slight focusing differences between the left and right eyes.

☐ **3** The **body tube** holds the ocular in place. Although called a "tube," it may be more like a box in some models.

☐ **4** At the bottom of the body tube is the **revolving nosepiece.** A turretlike circular mechanism rotates so that different lenses can be selected. *Always rotate the nosepiece by holding the outside of the revolving disk—never push on the lens barrels.*

☐ **5** Each of the lens sets attached to the revolving nosepiece is an **objective.** As with the ocular, each objective is marked with its magnification factor. Microscopes may have any or all of the following objectives:

4× (scanning objective) is used for initial location of the specimen.

10× (low-power objective) may also be used for initial location of the specimen. It is also used for observing specimens that don't need greater magnification.

Body tube

Arm

Mechanical stage clip

Mechanical stage adjustment

Coarse adjustment

Fine adjustment

Base

Ocular lenses

Ocular adjustment

Nosepiece

Objective lenses

Condenser (below stage)

Stage

Iris diaphragm

Condenser (movable)

Light source

Light intensity dial

Light switch

Figure 1.1 Important parts of a typical compound light microscope.

40× (high-dry objective) is used for specimens requiring greater magnification. This objective is called *dry* because it does not require the use of oil, as other high-power objectives do. Some scopes have 43× or 45× high-dry objectives. List the factor on your scope's high-dry objective here: _____.

100× (high-oil objective) is used for magnification of extremely small specimens, such as bacterial cells. It must be immersed in oil, so it is called the high-oil objective. This objective will not be used in this course.

In this manual, *low power* will refer to use of the 10× objective and *high power* will refer to use of the 40× objective.

☐ **6 Total magnification** is determined by multiplying the power of the ocular by the power of the objective in use. Thus, when using a 10× ocular and the low-power objective, total magnification is 100×:

$$10 \times 10 = 100$$

Likewise, when using a 10× ocular and a 40× objective, total magnification is 400×. List all ocular and objective combinations, then determine all the total magnifications possible on your scope and record them in the lab report.

☐ **7** The specimen is usually mounted on a glass or plastic **microscope slide** that rests on the **stage,** a platform just below the objective. The stage has a hole so that light can pass through the specimen from below. If the stage has an adjustable bracket that moves the slide around mechanically by rotating **positioning knobs,** the stage is called a **mechanical stage.** If not, the slide is held by **stage clips** and must be moved by hand.

☐ **8** Below the stage is a high-intensity **lamp.** Light rays from the lamp travel through a hole in the stage, through the specimen mounted on a slide, then through the objective and ocular, to the eye.

☐ **9** A **condenser,** a lens that concentrates light, may be found between the lamp and the stage.

☐ **10** Sometimes the light from the lamp is too strong to see the specimen clearly. The light level may be reduced by adjusting the lamp intensity (if possible). Light intensity may also be adjusted by adjusting the **diaphragm** just below the stage (fig. 1.2). A *disk diaphragm* is a rotating disk with holes of different diameters. An *iris diaphragm* is made of overlapping slivers of metal in a circular pattern resembling the iris of the human eye. As in the iris of the eye, the iris diaphragm can dilate or constrict its opening. You can change the amount of light passing to the specimen by rotating the edge of the disk diaphragm or by rotating the lever projecting from the iris diaphragm.

(a)

(b)

Figure 1.2 (*a*) Disk diaphragm set at a large opening, then at a smaller opening. (*b*) Iris diaphragm open, then partially closed.

☐ **11** The entire upper assembly of the microscope is held in an upright position by a bar called the **arm.** The scope is supported by a square or horseshoe-shaped **base.** The arm may be connected to the base by a **pivot,** which allows the upper assembly to move into a more comfortable viewing position.

☐ **12** The **coarse-focus knobs** and **fine-focus knobs** are on the arm. These knobs adjust the distance between the stage and objective, thus focusing an image of the specimen. The fine-focus knob changes the distance very little, whereas the coarse-focus knob changes the distance greatly.

B. Using the Microscope

Skill in using the microscope is necessary for many of the exercises in this lab manual. Fortunately, learning to use the microscope is both easy and fun. Before beginning, acquaint yourself with these basic rules.

○ Always carry the scope with two hands: one under the base and the other grasping the arm. Carry it in an upright position.

○ Carefully unwind the lamp cord. Avoid damaging the parts around which it is coiled. Plug the cord into an outlet in a safe manner.

SAFETY FIRST!

Be careful when plugging the power cord into the outlet. Always plug it into a receptacle at *your* station, taking care not to string it over a chair, across an aisle, or in any other dangerous position.

○ Make sure the stage and objective are at their farthest distance apart and that the *lowest-power objective is in position.* Start each new observation at low power.

☐ **1** Obtain a microscope slide with a newsprint letter *e* mounted on it. This is your practice specimen.

☐ **2** Carefully place your slide on the stage and secure it with stage clips or the brackets of the mechanical stage. Move the slide so that the *e* is centered in the stage's hole.

☐ **3** While looking from the side, use the coarse-focus knob to move the objective (still on low power) as close as possible to the slide.

☐ **4** Look through the ocular and use the coarse-focus knob to slowly move the objective and slide apart. When the image becomes clear, switch to the fine-focus knob to make the image even sharper. To avoid damaging the scope and slide, *never move the objective and stage toward each other while looking through the ocular.*

Hint . . . To avoid eyestrain and a possible headache, always keep both eyes open when viewing a specimen. This is easy with binocular scopes, once you have adjusted the distance between oculars. However, it may take some practice if you are using a monocular scope. If you have trouble keeping both eyes open, try covering the unused eye with your hand.

☐ **5** Adjust the light intensity by using the lamp controller or diaphragm until the detail of the image is at its clearest.

☐ **6** Sketch the entire field of view in the space provided in the lab report. Notice that the *e* appears upside down and backwards in your field of view. This inversion is caused by the way the lenses in the scope bend light to magnify the image. You will notice the same effect if you hold an ordinary magnifying glass at arm's length.

Hint . . . Here are some rules for properly recording microscopic observations:

• Label each sketch with the name of the specimen and what type of section it is.

• Label each sketch with the *total magnification.*

• Label as many parts of the sketch as you can. Labels should be orderly, never crossing lines with each other.

• Sketches should be done with care, not with haste.

☐ **7** Center the part of the specimen that you wish to see more clearly and switch to a higher-power objective. Notice that as you move the slide to the left, it *appears* to move to the right. This reversal is caused by the lenses in the scope. Most scopes are **parfocal,** which means that the image remains focused when you change objectives. However, some minor adjustment with the fine focus is usually necessary. *Never adjust the coarse focus when using high power—you are likely to break the slide and scratch the objective lens.*

☐ **8** Sketch the entire field that you see using the high-power (40× to 45×) objective.

☐ **9** When you remove the slide, make sure that the low-power objective is in viewing position and that the objective and stage are as far apart as possible.

☐ **10** Practice focusing at different depths by using a prepared slide of three crossed strands of colored thread. The strands (one red, one yellow, and one blue) cross at the same point. Determine which is on top, which is in the middle, and which is on the bottom. Report your results in the lab report.

☐ **11** When you have finished for the day, return the scope to its original configuration. The low-power objective should be in position. Make sure the slide has been removed.

Hint . . . If you can't see an image in your scope . . .

- Make sure the lamp is on.
- Make sure the diaphragm is open. The amount of light should be set low on low power and high on high power.
- Check for obstructions in the light path.
- Make sure the objective is seated properly.
- Check to see if the specimen is centered.
- Clean the lenses with lens paper and lens solution, not facial tissues or lab wipes.

If the image seems to fade in and out . . .

- Watch to see if the body tube or stage is shifting or dropping.
- Make sure the scope doesn't require a rubber eyecup for proper viewing.
- Make sure the specimen isn't in a medium that obstructs viewing.
- Check the lamp or cord for short circuits.

If the image doesn't look like the figure in the book . . .

- Get a different book (it may have a better figure).
- Use your imagination.
- Join the club (nobody else's specimen is identical to that in the book, either).

C. Preparing Microscopic Specimens

You will often use prepared slides of human and other tissues in this course. However, you may occasionally make your own specimens.

A **wet-mount** slide is a slide on which a wet specimen is placed, then covered with a **coverslip** (fig. 1.3).

Stains are used to make a specimen, or some of its parts, more visible. Some require special techniques, but most stains can simply be added to the specimen and viewed.

☐ **1** Obtain some skin cells by scraping the inner surface of your cheek with a clean, flat toothpick.

☐ **2** Wipe the scrapings on a clean microscope slide and put a small drop of *methylene blue* stain directly on the smear.

SAFETY FIRST!

Use only fresh toothpicks and dispose of used toothpicks *immediately* by placing them in the waste container provided by your instructor. Take precautions to avoid the spread of disease through saliva. Consider using gloves and protective eyewear.

☐ **3** Place one edge of a coverslip on the slide next to the specimen, let it touch the drop of liquid, then let it drop slowly on the specimen. This method avoids forming air bubbles.

☐ **4** Absorb any excess fluid around the edges of the coverslip with the edge of a paper wipe.

☐ **5** Locate some cells with the low-power objective, shift to high power, and sketch your observations in the lab report.

Landmark Characteristics

You are looking for scattered epithelial cells. Each cell will appear as a lightly stained, flat polygon or circle with a dark center. The dark center is the nucleus of the cell. Some cells are separate, whereas others are clumped together.

You may see a variety of *other things* in your specimen: bacteria, vegetable or meat fibers, and other matter. Don't worry; this is normal. Distinct dark circles with hollow centers that may be in motion are air bubbles.

Hint . . . If your instructor wants you to learn how to use an oil-immersion objective, follow these directions. After focusing on high dry, rotate the nosepiece *halfway* to the high-oil objective. Put a drop of immersion oil on the slide; then rotate the high-oil objective into the oil drop. You may need to increase the lighting. When finished, clean the objective and other surfaces as your instructor directs.

Figure 1.3   Staining a wet-mount specimen. (*a*) Place the specimen on the slide. (*b*) Add a small drop of stain. (*c*) Drop the coverslip slowly to avoid air bubbles. (*d*) Absorb excess fluid with a paper wipe.

LAB REPORT 1

The Microscope

Name _____

Date _____ Section _____

Specimen: *newsprint* e   Total magnification: _____

Which of the three colored threads in the prepared slide is on the bottom?

Which thread is on the top?

Specimen: *human cheek cells*   Total magnification: _____

What is the darkly stained region of each cell?

What advantage is gained by staining this specimen?

Specimen: *newsprint* e   Total magnification: _____

Does the *e* in the newsprint appear to be oriented the same way it is on the stage? Explain your observations.

Is the *diameter of field* the same when observing the specimen under high power and low power?

List all the possible ocular-objective combinations on your microscope and calculate the total magnification for each.

Power of Ocular	Power of Objective	Total Magnification
_____	_____	_____
_____	_____	_____
_____	_____	_____
_____	_____	_____

Fill-in

_____ 1

_____ 2

_____ 3

_____ 4

_____ 5

_____ 6

_____ 7

_____ 8

Multiple choice

_____ 1

_____ 2

_____ 3

_____ 4

_____ 5

_____ 6

_____ 7

Fill-in

1. A(n) __?__ is a glass or plastic rectangle on which specimens are mounted.
2. A(n) __?__ is at the base of the body tube, allowing selection of different objectives.
3. Below the stage is a high-intensity __?__ or another light source.
4. Different lens sets, or __?__, are attached to the revolving nosepiece.
5. If the stage has a bracket that moves the slide, it is called a(n) __?__ stage.
6. The __?__ allows the user to limit the amount of light from a steady light source.
7. The set of lenses closest to the viewer's eye is the __?__.
8. The __?__ is a lens under the stage that concentrates light.

Multiple choice

1. A microscope __?__ the image of a specimen.
   a. magnifies
   b. reverses
   c. reduces
   d. erases
   e. *a* and *b*

2. The use of __?__ may make the specimen (or some of its parts) more visible.
   a. stain
   b. the diaphragm
   c. fine focus
   d. all of the above
   e. none of the above

3. If you select a 40× objective to use with a 5× ocular, the total magnification is
   a. 200×.
   b. 45×.
   c. 400×.
   d. 405×.
   e. none of the above.

4. When you move a specimen to the right, its image appears to
   a. move to the right.
   b. move to the left.
   c. remain stationary.

5. Which is the correct path of light in a compound microscope?
   a. lamp, objective, diaphragm, specimen, ocular
   b. lamp, diaphragm, specimen, objective, ocular
   c. diaphragm, lamp, specimen, ocular, objective
   d. lamp, diaphragm, specimen, ocular, objective

6. If your scope is not working properly, consult
   a. the lab instructor.
   b. your lab partner.
   c. the owner's manual.
   d. the manufacturer.

7. When carrying your microscope, use __?__ hand(s).
   a. one
   b. two

Cell Anatomy

The cell is often considered the basic unit of living organisms. Actually, cells are only one of several *levels of organization* in the human organism:

chemicals
organelles
cells
tissues
organs
organ systems
organism

Thus, cells are parts of larger units (tissues and organs) and are composed of smaller units (organelles).

For now, we'll concentrate on the structure and function of the cell and its organelles. We will study some of the other levels of organization in later units.

Before You Begin

☐ Read the appropriate chapter in your textbook.

☐ Set your learning goals. When you finish this exercise, you should be able to:
  o place the cell and its organelles within the scheme of organizational levels
  o identify the major organelles on a model, chart, specimen, or micrograph
  o describe primary function(s) of typical organelles
  o understand the anatomical relationship among various cell parts

☐ Prepare your materials:
  o cell model or chart
  o compound light microscope
  o clean slides
  o coverslips
  o paper wipes
  o methylene blue stain
  o colored pencils or pens
  o prepared slides—instructor's option
  o glove
  o protective eyewear

☐ **Carefully** read the directions and safety tips for this exercise before starting any procedure.

A. Parts of the Cell

A cell is a microscopic unit made up of a bubble of fatty material filled with a water-based mixture of molecules and tiny particles. Parts of any cell are called **organelles** (meaning *small organs*). Begin your study of the cell by examining a *generalized* cell. A generalized cell is one that has many cell features that are not all found in a single natural cell. Locate each of the listed organelles on a generalized cell model or chart (fig. 2.1).

☐ **1** The outer boundary of the cell is the **cell membrane,** or **plasma membrane.** It is composed of a double layer (or *bilayer*) of phospholipid molecules embedded with other molecules. The cell membrane has many functions, most involving transport and communication between the inside and outside of the cell.

☐ **2** The material enclosed by the cell membrane is called **cytoplasm.** Cytoplasm includes both the organelles and the liquid inside the cell.

Hint . . . Each membranous organelle described below is made of a lipid membrane similar to the cell membrane. Because they are so thin, cell membranes may seem invisible when observed with a light microscope.

☐ **3** The double-walled **nucleus** is a large bubble containing the cell's genetic code. The code is in the form of **deoxyribonucleic acid (DNA)** located in strands called **chromatin.** Portions of chromatin accept stains readily, giving the nucleus a very dark appearance.

☐ **4** The **nucleolus,** literally *tiny nucleus,* is a small area within the nucleus for the synthesis of ribosomal **ribonucleic acid (RNA).**

☐ **5** The **endoplasmic reticulum (ER)** is a network of membranous tubes and canals winding through the interior of the cell. A *rough ER* is specked with tiny granules (ribosomes); a *smooth ER* is not. The ER transports proteins synthesized by ribosomes and other molecules synthesized within its membrane. ER also manufactures molecules that make up cellular membranes.

The Cell

 Figure 2.1 Use colored pens or pencils to shade in the cell parts shown and their corresponding labels. Use contrasting colors for different parts.

□ **6** **Ribosomes** are tiny bodies that serve as sites for protein synthesis. Some ribosomes are found on the outer surface of ER, and some are found scattered elsewhere within the cell.

□ **7** The **Golgi apparatus** appears as a stack of flattened sacs. The apparatus receives material from the ER, processes it, then packages it in tiny *vesicles* (bubbles) for possible export from the cell.

□ **8** **Mitochondria** (sing., **mitochondrion**) are tiny bodies similar to bacteria that serve as sites for ATP synthesis (energy conversion). Mitochondria have an outer membrane, forming a round or oblong capsule, and a folded inner membrane. The folds of the inner membrane are called *cristae.*

□ **9** **Lysosomes** are vesicles containing digestive enzymes that digest foreign particles and worn cell parts.

□ **10** **Microtubules** are very tiny, hollow beams that form part of the supporting cell skeleton, or *cytoskeleton.* They also form parts of other cell organelles, such as *flagella, cilia, centrioles,* and *spindle fibers.* Other components of the cytoskeleton include **microfilaments** and **intermediate filaments.** The cytoskeleton, as the skeleton of the body, gives shape to the cell, provides the mechanisms for movement, and provides attachment points for organelles.

□ **11** The **centrosome,** or *microtubule organizing center,* is a dense area of cell fluid near the nucleus. The centrosome contains a pair of **centrioles,** cylinders formed by parallel microtubules. A network of microtubules called *spindle fibers* extends from the centrosome during cell division. Spindle fibers distribute DNA equally to the resulting daughter cells.

□ **12** The cell may have any number of other assorted organelles. **Microvilli** are tiny, fingerlike projections of the cell that increase the membrane's surface area for more efficient absorption. **Cilia** are numerous short, hairlike organelles that propel material along a cell's surface. **Flagella** are single long, hairlike organelles found in sperm cells to propel them through the female reproductive tract toward the egg. **Vesicles** are membranous bubbles that may be formed by the Golgi apparatus or by the pinching inward of the cell membrane to engulf external substances.

B. Microscopic Cell Specimen

Prepare a stained wet-mount specimen of human cheek cells (see Lab Exercise 1 for instructions) or obtain another specimen provided by the instructor; try to identify as many cell parts as possible. Use high-power magnification (after first finding the specimen with low-power magnification). Sketch and label your observations in the lab report.

Hint . . . Many organelles are very small and can be seen only when properly prepared and examined with a more powerful microscope.

C. Interpreting Micrographs

An **electron microscope** is an instrument that uses a beam of electrons, rather than a beam of light, to form the image of a tiny specimen. *Transmission electron microscopes* send an electron beam through the specimen, similar to the manner in which a light microscope sends a light beam through a specimen. However, the magnifying power and resolution are much greater in the electron microscope. **Resolution** is the ability to distinguish detail. *Scanning electron microscopes,* on the other hand, reflect an electron beam off

the specimen. The shadows produced by a scanning electron beam lend a three-dimensional effect.

A **TEM** is a transmission electron micrograph (photograph taken with a transmission electron microscope). Figure 2.2 shows TEM representations of major cell parts. Try to identify all of them by completing the labeling exercise. This will not be very easy. Trying to identify structures in the TEM, however, will be a worthwhile challenge. You will clearly see how difficult it is to apply general concepts to individual specimens.

1

2

3

4

5

Figure 2.2   Identify the cell structures in this TEM of human epithelial cells (×7900) (© EM Research Services, Newcastle University/RF).

LAB REPORT 2

Cell Anatomy

Name _____

Date _____ Section _____

Checklist for Cell Model

☐ cell membrane

☐ nucleus

   ☐ chromatin

☐ nucleolus

☐ endoplasmic reticulum

   ☐ rough ER

   ☐ smooth ER

☐ ribosomes

   ☐ free

   ☐ associated with ER

☐ Golgi apparatus

   ☐ secretory vesicles

☐ mitochondria

☐ lysosomes

☐ microtubules

☐ centrosome

   ☐ centrioles

☐ microvilli

☐ cilia

☐ flagella

☐ _____

☐ _____

☐ _____

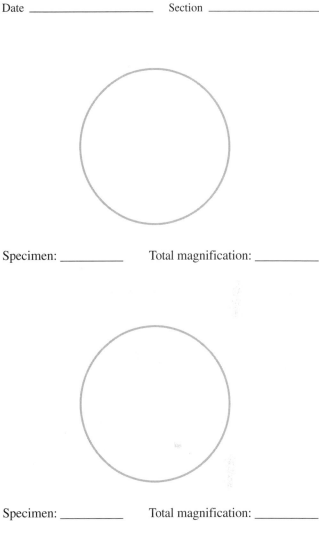

Specimen: _____ Total magnification: _____

Specimen: _____ Total magnification: _____

Specimen: _____ Total magnification: _____

2-6

Figure 2.2 *(from p. 12)*

_____ 1

_____ 2

_____ 3

_____ 4

_____ 5

Matching

_____ 1

_____ 2

_____ 3

_____ 4

_____ 5

_____ 6

_____ 7

_____ 8

_____ 9

_____ 10

Fill-in

_____ 1

_____ 2

_____ 3

_____ 4

_____ 5

Matching

Answer may be used more than once or not at all.

a. cell membrane
b. centriole
c. endoplasmic reticulum
d. Golgi apparatus
e. lysosome
f. mitochondrion
g. nucleolus
h. nucleus
i. ribosome

1. A double-walled structure containing the cell's genetic code
2. A network of membranous tubes and canals that transports proteins
3. A stack of flattened sacs that process and package proteins
4. The site of manufacture of ribosomal RNA
5. A cylinder formed by parallel microtubules
6. An organelle that serves as the site of protein synthesis
7. A bubble containing digestive enzymes
8. May be "rough" (with ribosomes) or "smooth" (ribosome-free)
9. Allows communication between the internal and external cell environment
10. Forms secretory vesicles

Fill-in table

Write the names of the organelles listed in the Matching section (*a–i*) in the appropriate column of the table.

Membranous	Nonmembranous

Fill-in paragraph

The cell membrane is a bilayer made of __1__ studded with protein. The membrane's __2__ molecules prevent water-soluble molecules from entering, whereas the __3__ molecules serve as channels for specific molecules. The membrane's protein molecules may also serve as anchors for __4__ that form part of the cytoskeleton. The same material forms the primary structural elements of __5__ organelles, such as the mitochondrion and ER.

Transport through Cell Membranes

Cells make use of both **passive** and **active** processes to transport substances across its membranes. Passive processes are those that require no metabolic energy (ATP) from the cell but rely solely on the physical properties of the substances themselves. Active transport processes require energy (ATP) expenditure by the cell to move substances whose physical properties prevent their moving on their own. In this exercise, we will observe large-scale examples of passive transport processes.

Before You Begin

☐ Read the appropriate chapter in your textbook.

☐ Set your learning goals. When you finish this exercise, you should be able to:
  o understand the nature of diffusion and osmosis
  o use the terms *hypertonic, hypotonic,* and *isotonic* to compare solutions
  o compare and contrast diffusion, osmosis, and filtration

☐ Prepare your materials:
  o India ink (dropper bottle)
  o microscope
  o microscope slides and coverslips
  o potassium permanganate crystals
  o Petri dishes
  o hot (100°C) water and cold (0°C) water
  o forceps
  o dialysis tubing (8 to 10 cm, presoaked)
  o string (or tubing clamps)
  o sucrose solutions (15% and 40%)
  o syringe (or pipette) to fill dialysis bags
  o laboratory balance
  o jar or beaker (75 to 200 mL)
  o distilled water
  o filter paper (circular, large pore)
  o glass funnel
  o flask (to hold funnel and filtrate)
  o copper sulfate (saturated solution)
  o boiled starch solution (10%)
  o charcoal (powdered)
  o Lugol's reagent (iodine solution)
  o PHYSIOLOGY INTERACTIVE LAB SIMULATIONS (PH.I.L.S.) 4.0 and computer system (optional)

☐ **Carefully** read the directions and safety tips for this exercise before starting any procedure.

A. Physical Basis of Passive Transport

Naturally occurring **Brownian motion** drives passive transport processes. Discovered by Scottish scientist Robert Brown, Brownian motion is the constant movement of all particles of matter.

☐ **1** Prepare a wet-mount slide with a drop of India ink.
☐ **2** Observe the slide under high-power magnification. You should be able to see the dye particles moving short, irregular distances. This is Brownian motion.

Because small particles of matter are always bouncing around, Brownian motion causes matter to spread, or **diffuse,** to areas where there is more room to bounce around. That is, particles tend to move toward an area of lower particle concentration. You can observe this movement by following steps 3 and 4.

☐ **3** Using forceps, place a small crystal of potassium permanganate in the center of a Petri dish of calm, hot water (fig. 3.1). Placing a sheet of clean, white paper under the dish will help you see the crystal more clearly. Place a second crystal of the same size in a dish of calm, cold water. Be careful not to disturb the dishes by touching them or causing them to shake.
☐ **4** Observe the two dishes over a period of 30 minutes, carefully noting any changes in the crystals. Answer the questions in the lab report.

B. Diffusion and Osmosis

Diffusion can be defined as the net movement of particles from an area of high concentration to an area of low concentration. Some particles move down such a *concentration gradient* when they are allowed to cross a cell membrane and some of the solute particles cannot cross the membrane freely (fig. 3.2). If the moving particles in question happen to be water molecules, the process is called **osmosis.** Osmosis, then, is a particular type of diffusion. In this activity, you will observe the osmosis of water across a nonliving membrane called *dialysis tubing.* Dialysis tubing, like the cell membrane, is *semipermeable* (allowing only certain substances to pass through its pores). The bag that

Figure 3.1 Slowly place the potassium permanganate crystal, taking care not to disturb the water.

you construct from a piece of dialysis tubing serves as a model of the cell (fig. 3.3).

One of the three model cells you construct will be immersed in an **isotonic** solution. *Isotonic* is a comparative term that refers to a solution that has the same *potential osmotic pressure* as the cell. In other words, the solutions inside and outside the cell have the same relative water concentrations. When two solutions have the same relative water concentration, there is no net movement of water and therefore no change in pressure.

The second cell model will be immersed in **hypertonic** solution, a solution with a higher potential osmotic pressure

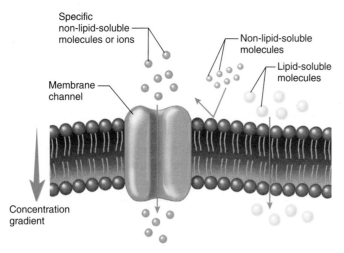

Figure 3.2 Diffusion through a cell membrane. Non-lipid-soluble molecules diffuse through membrane channels. Lipid-soluble molecules diffuse directly through the cell membrane.

Figure 3.3 A cell in (*a*) hypotonic solution, (*b*) isotonic solution, and (*c*) hypertonic solution. The arrows show the direction of net osmosis.

(lower water concentration). The third cell model will be immersed in **hypotonic** solution. A hypotonic solution is one with a lower potential osmotic pressure (that is, a higher water concentration) than the cell. When the *tonicity* of a solution outside a cell's plasma membrane is known, the net direction of osmosis can be predicted.

Hints . . . These hints may help you:

- Either presoak the dialysis tubing before using it or hold it under water and rub the end of the tube between your fingers until it opens.
- Set up the jars described in step 4 *before* beginning this activity.
- Because this activity calls for an extended waiting period, you may want to do the other activities of this exercise while you wait.

As you set up each of the three situations described below, predict the direction of net osmosis.

☐ **1** Obtain three 8 to 10 cm pieces of dialysis tubing. Tie one end shut with string, forming a water-tight bag. (Longer pieces of tubing can be tied by simply tying each end on itself.) Leave about 10 cm of string dangling freely.

☐ **2** With a syringe, carefully fill each of the three bags with 15% sucrose solution. Try to avoid spilling any solution on the outside of the bag. Tie off the open end of each tube with another piece of string.

☐ **3** Quickly measure the mass of each tube on a laboratory balance.

☐ **4** Place bag 1 in a jar or beaker half-filled with distilled water. Place bag 2 in a jar half-filled with 15% sucrose solution. Place bag 3 in a jar half-filled with 40% sucrose solution. Leave a piece of string dangling out of the jar, as you would with the string on a tea bag (fig. 3.4).

Intracellular

Extracellular

(a) (b)

Figure 3.4 (*a*) Folding the end of the dialysis tube before tying it prevents leaking. (*b*) Your model cell contains a sucrose solution that represents *intracellular* fluid. The bath solution represents *extracellular* fluid.

□ **5** After 45 to 60 minutes, remove the tubes from the jars, drying and weighing each as you do so. Record and interpret your results in Lab Report 3.

C. Filtration

The movement of particles across a membrane from an area of high pressure to an area of low pressure (down a *hydrostatic pressure gradient*) is called **filtration.** In this experiment, gravity creates the pressure needed to push some particles across a paper membrane. In the human body, forces such as blood pressure push particles through living membranes.

□ **1** Place a cone made of filter paper inside a glass funnel.

Hint . . . To make the cone, fold a circle of paper in half, then in half again. You should now have a triangle four layers thick. Pull one of the layers away from the rest. You now have a paper cone. Place it in the funnel and wet it with distilled water so that it sticks against the side of the funnel.

□ **2** Position the funnel over an empty flask, as shown in figure 3.5.
□ **3** Mix together 3 mL of copper sulfate solution, 4 mL of boiled starch solution, a pinch of charcoal powder, and 5 to 8 mL of distilled water. Note the physical characteristics of each as you mix them.
□ **4** Pour the mixture into the funnel.
□ **5** Observe the mixture being filtered by the paper. Does the process become progressively faster, become progressively slower, or remain steady? Record and explain your results in Lab Report 3.

□ **6** Examine the *filtrate* (filtered material). Test for the presence of each component of the original mixture:
a. *charcoal*—observe visually
b. *copper sulfate*—observe visually
c. *starch*—add a few drops of Lugol's reagent, or iodine solution (a color change to black means that starch is present)

What substances were not allowed through the filter? Record and explain your results in Lab Report 3.

Funnel

Paper filter

Flask

Figure 3.5 Set up your filtration apparatus as shown.

Clinical Application

Results similar to those in your osmosis experiment occur when red blood cells come into contact with various solutions that are injected into the blood supply, as in *intravenous (IV) therapy*. When blood cells are bathed in a solution that is isotonic to them, they remain unchanged (fig. 3.6). If a hypotonic solution is introduced, the cells experience an inflow of water and usually burst. Bursting of red blood cells caused by osmosis is called **hemolysis.** If hypertonic solution is introduced, the cells lose water and shrivel. Shriveling that results from osmotic loss of water is called **crenation.** Therefore, the concentration of injected materials is critical to the survival of the patient.

Figure 3.6 The effects on red blood cells when changing the tonicity of the extracellular fluid (plasma).

Predict the expected results in these situations:

☐ **1** Randy is a nursing student. He has been asked to intravenously inject his patient with 10 cc of an isotonic preparation. Mistakenly, he injects his patient with 10 cc of pure water. What is likely to happen to the red blood cells near the site of injection?

☐ **2** Jennifer is Randy's patient. She is carefully watching him fill an IV bottle with a mixture specially ordered by the physician. She notices that he accidentally fills the bottle with concentrated (10%) salt solution instead of the mixture from the pharmacy. Why should she refuse to allow Randy to attach the IV bottle to her system?

☐ **3** Randy is now assigned to the surgical unit. He is given a piece of living tissue and asked to put it in fluid before taking the sample to the pathology lab. What essential characteristic must such fluid have so that it will not damage any cells in the tissue?

Transport through Cell Membranes

Name _____

Date _____ Section _____

1. Describe the movement of India ink particles. Was it uniform or random? Were the particles moving great distances?

Observation	Elapsed Time	Extent of Diffusion in COLD WATER	Extent of Diffusion in HOT WATER
1	*0 minutes*	*No diffusion of dye particles*	*No diffusion of dye particles*
2			
3			
4	*30 minutes*		
Results			

2. Using potassium permanganate as an example, explain in your own words how Brownian motion relates to diffusion.

3. Based on your results in the potassium permanganate experiment, would you state that increased temperature increases or decreases Brownian motion?

Bag	Predicted Result	Initial Mass	Final Mass	Gain or Loss of Mass	Explanation of Results
1					
2					
3					

Advanced Scientific Inquiry
PHYSIOLOGY INTERACTIVE LAB SIMULATION (PH.I.L.S. 4.0)

If available, use the computerized laboratory simulation PH.I.L.S. 4.0 to simulate the following activity that explore cell transport function:

For each of the activities you perform, print the lab report according to the directions given in the activity and attach a copy of all sheets to your lab report (or as your instructor directs).

- *Varying Extracellular Concentration (simulation #1)*

4. Did the filtration rate of your mixture become progressively faster or slower, or was it steady? Explain.

Substance	Test Result (Present or Absent)	Explanation of Result
Charcoal		
Copper sulfate		
Starch		

Matching

_____ 1

_____ 2

_____ 3

_____ 4

_____ 5

_____ 6

_____ 7

_____ 8

_____ 9

_____ 10

Identify

_____ 1

_____ 2

_____ 3

_____ 4

_____ 5

Matching

Answer may be used more than once or not at all.

a. Brownian motion
b. diffusion
c. filtration
d. hydrostatic pressure
e. hypertonic
f. hypotonic
g. isotonic
h. osmosis
i. semipermeable

1. The tendency of matter to spread to areas of lower concentration
2. Movement through a membrane driven by a hydrostatic pressure gradient
3. The term that specifically describes the diffusion of water across a membrane
4. The term that describes a membrane that allows only some types of particles to pass through it
5. The natural vibration of particles; it drives diffusion
6. In the kidney, blood pressure forces some water and solute particles from a blood vessel and into a kidney tubule. What is this type of transport called?
7. A cell is bathed in solution X. The cell quickly shrivels. What term describes solution X (when compared with the cell's fluid)?
8. Solution Q has a higher water content than cell Z. Therefore, solution Q is __?__ to cell Z.
9. A saline (salt) solution is to be injected into a patient. The salt/water ratio should be adjusted so that the saline solution is __?__ to the patient's cells.
10. Particles of substance Y move into a cell because there are fewer particles inside the cell than outside. This is an example of __?__.

Identify

State whether each item is an example of *diffusion, filtration,* or *active transport.*

1. Movement of water from an area of high concentration to an area of low concentration
2. Dye particles spread evenly through water.
3. Starch particles pass through a paper membrane.
4. A cell using energy to "pump" sugar molecules from its external environment
5. Osmosis

The Cell's Life Cycle

Cells in many parts of the human body divide to produce more cells of the same type. The hereditary information contained within the nucleus of a resting *parent cell* must first be **replicated** (copied), then evenly distributed between the two cells that result from division. The process of distributing genetic material is termed **mitosis.** Mitosis was named in the late nineteenth century by Walther Flemming, who noticed threadlike structures in cells during cell division (*mitos-,* "threads"; *-osis,* "condition of").

The pinching in of the cell membrane, and eventual split of the membrane and its contents into two *daughter cells,* is termed **cytokinesis.** Cytokinesis occurs about the same time (or just after) the last phases of mitosis.

Life cycles are circular patterns of organisms' life histories. For example, the life cycle of humans includes conception, development, adulthood, reproduction, then conception and so forth again for the offspring. Likewise, individual body cells are formed, they reproduce, and their daughter cells continue the cycle of life. In this exercise, we will explore the major events of the human cell's life cycle.

Before You Begin

☐ Read the appropriate chapter in your textbook.

☐ Set your learning goals. When you finish this exercise, you should be able to:
  o list the major phases of a cell's life cycle
  o describe the principal events of mitosis
  o identify the cell parts involved in mitosis
  o explain the importance of mitosis
  o define the term *cytokinesis*

☐ Prepare your materials:
  o chart or model: *animal mitosis series*
  o colored pencils or pens
  o microscope
  o prepared microslide: *whitefish blastula*

☐ **Carefully** read the directions and safety tips for this exercise before starting any procedure.

A. Introduction to the Cell Life Cycle

The process of mitosis can be described step by step to make it a little easier to picture. We will divide the whole life cycle into five phases: **interphase, prophase, metaphase, anaphase,** and **telophase** (fig. 4.1).

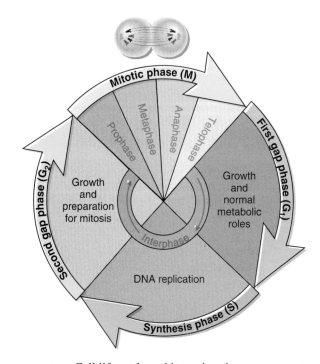

Figure 4.1   Cell life cycle and its major phases.

Note the major events of each life cycle phase listed. Try to identify the physical representation of those events in a chart or model of cells at various life cycle stages.

☐ **1**   Interphase is not a phase of mitosis but is the period between cell divisions. It is not an inactive time, however, since chromatin (DNA) replicates during interphase (forming two sister **chromatids** joined at a **centromere**). In anticipation of division, additional cell fluid and organelles are formed during interphase.

Hint . . . The cell life cycle is a continuous process. The cell doesn't suddenly jump from one phase to another but gradually changes. Therefore, cell models represent a snapshot of each phase at a point where it is most distinct from the phases before and after it.

☐ **2** Prophase is the first phase of mitosis. During prophase, the nuclear membrane disappears, freeing the chromatin (which first shortens into tiny bodies called **chromosomes**). Also, centriole pairs move to opposite poles of the cell as **spindle fibers** begin to project from them. The spindle fibers extend toward the equator of the cell and may overlap with spindle fibers projecting from the opposite centriole pair. A spindle fiber may also attach to one side of a chromosome's centromere.

☐ **3** Metaphase is the period during which the chromosomes (each a pair of replicate chromatids joined at a centromere) arrange themselves singly as a thin sheet along the cell's equator (imaginary center plane). Each chromosome now has a pair of spindle fibers attached to it, one from each centriole pair.

☐ **4** Anaphase is the phase during which the chromatids split at the centromere, each moving toward an opposite pole along the path of a spindle fiber. At the end of anaphase, each pole of the cell has a full group of 46 single chromosomes. The chromosomes on one side of the cell are replicates of the set of chromosomes on the other side.

☐ **5** Telophase is the time during which each side of the cell changes everything to the way it should be during interphase:
  o A nuclear membrane forms around each group of chromosomes.
  o The chromosomes uncoil to form long chromatin strands.
  o Remnants of the spindle fibers disintegrate.

☐ **6** During anaphase of mitosis, the separate (but concurrent) process of cytokinesis begins. By the end of anaphase, cleavage, or pinching in, of the parent cell is evident. By the end of telophase, complete splitting of the parent cell into two similar daughter cells is complete. Each daughter cell has a nucleus and roughly half of the cytoplasm and organelles of the parent cell.

☐ **7** Review the major events of the cell life cycle by completing the coloring exercise "Cell Life Cycle" (fig. 4.2).

The length of time between divisions, and the time required for division to take place, varies considerably from cell to cell. Even the relative length of different phases varies among individual cells. Cell division can range from 20 minutes to several hours.

B. Microscopic Observations

In this activity, you will observe cells of the whitefish blastula in a prepared slide (fig. 4.3). The blastula is a developmental stage in the growth of many animals. The original cell formed by the joining of the egg and sperm undergoes mitosis many times, at one stage forming a ball of cells called a *blastula*. The cells of the blastula are still undergoing rapid cell division, making it a specimen likely to have many cells at various stages of the cell life cycle.

SAFETY FIRST!

Don't forget to check the microscope's power cord for frays and for proper placement. Take care that you don't crack a slide or lens by zooming in with coarse focus without looking from the side. Accidents are less likely to happen when you are careful.

☐ **1** Obtain a prepared microscope slide (microslide) of whitefish blastula cross sections (c.s.).

☐ **2** Locate a group of cells using low-power magnification. There are usually several different cell groups from which to choose on each slide.

☐ **3** Switch to high power and try to locate individual cells in each of the five phases of the cell life cycle. You may need to switch slides occasionally to find all the phases.

Landmark Characteristics

Many prepared whitefish blastula specimens are stained so that the cytoplasm appears pinkish to brown-red (see Plate 1 in The Histology Minireference). The DNA (chromatin, chromosomes) is stained black. Therefore, look for fine, black formations when looking for the major events of mitosis. The cell membranes, spindle fibers, and centrioles are so small that they will not appear distinctly in your specimen.

In a single microscopic field, you will see many cells, all at different points in the cell life cycle. There is no pattern to the way the cells in different phases are arranged (they are not laid out in order of mitotic phases, for instance). If you are fortunate, you may find at least one example of each phase in a single field.

Cell Life Cycle

MITOSIS ₁

CYTOKINESIS ₂

INTERPHASE ₃

PROPHASE ₄

METAPHASE ₅

ANAPHASE ₆

TELOPHASE ₇

CELL (PLASMA) MEMBRANE ₈

NUCLEAR MEMBRANE ₉

NUCLEOLUS ₁₀

CENTRIOLES ₁₁

CHROMATIN/CHROMOSOME ₁₂

CENTROMERE ₁₃

Figure 4.2   Use colored pens or pencils to shade these structures and labels.

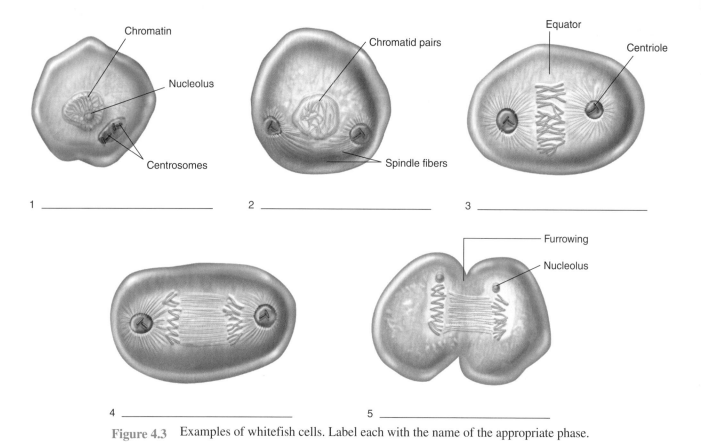

Figure 4.3   Examples of whitefish cells. Label each with the name of the appropriate phase.

1 _____   2 _____   3 _____

4 _____   5 _____

Another Kind of Cell Division

Meiosis is a process distinct from mitosis. Meiosis is associated with a type of cell division that occurs during the formation of reproductive cells (*sperm* and *eggs*). Mitosis occurs in the division of all other cell types.

Meiosis results in daughter cells that have only half the number of chromosomes that other cells, including the parent cells, have. That is, during meiosis, a parent cell with 46 chromosomes produces daughter cells that have 23 chromosomes in each. This must occur so that, when the sperm and egg unite during conception, the newly formed cell has 46 chromosomes (23 from the sperm plus 23 from the egg). Thus, the offspring has equal amounts of hereditary information from each parent.

Theoretically, what would happen if meiosis did not occur and sperm and egg cells could only form using mitosis?

What would happen if *all* your body cells divided using meiosis instead of mitosis?

LAB REPORT 4

The Cell's Life Cycle

Name _____

Date _____   Section _____

Sketch your observations of particularly clear examples of mitotic phases in whitefish cells.

Specimen: *whitefish, phase* _____
    Total magnification: _____

Specimen: *whitefish, phase* _____
    Total magnification: _____

Specimen: *whitefish, phase* _____
    Total magnification: _____

Specimen: *whitefish, phase* _____
    Total magnification: _____

Specimen: *whitefish, phase* _____
    Total magnification: _____

Specimen: *whitefish, phase* _____
    Total magnification: _____

Figure 4.3 (*from p. 24*)

_____ 1

_____ 2

_____ 3

_____ 4

_____ 5

Fill-in

_____ 1

_____ 2

_____ 3

_____ 4

_____ 5

_____ 6

_____ 7

_____ 8

_____ 9

_____ 10

Multiple choice

_____ 1

_____ 2

_____ 3

_____ 4

_____ 5

Fill-in

Provide the name of each mitotic phase.

1. The centrioles move toward opposite poles during ___?___.
2. During ___?___ the nuclear membrane disintegrates.
3. The nuclear membrane reappears during ___?___.
4. The last phase of mitosis is ___?___.
5. During ___?___ the chromosomes align at the cell's equator.
6. Cytokinesis usually begins during ___?___ of mitosis.
7. The first phase of mitosis is ___?___.
8. During ___?___ the chromatids separate and move toward opposite poles.
9. Spindle fibers appear during ___?___.
10. Mitosis ends with ___?___.

Multiple choice

1. The function of mitosis is
    a. to distribute the cell's DNA equally between the daughters.
    b. to divide the cytoplasm equally between the daughter cells.
    c. to distribute the parent cell's organelles evenly.
    d. all of the above
    e. none of the above

2. The chromosomes of daughter cells formed during mitosis
    a. are double the number of those in the parent cell.
    b. are identical to the chromosomes of the other daughter.
    c. are half the number of those in the parent cell.
    d. *b* and *c*
    e. all

3. Which of these cells is likely to have been formed using mitosis?
    a. skin cell
    b. sperm cell
    c. egg cell
    d. heart muscle cell
    e. *a* and *d*

4. A certain stain colors DNA violet. When stained, which of these should appear violet?
    a. chromatin
    b. chromosomes
    c. chromatids
    d. *a* and *b*
    e. all of the above

5. Two DNA molecules, joined at a centromere, have coiled to form tiny bodies. This connected pair of DNA molecules should properly be called a
    a. centromere.
    b. centriole.
    c. chromosome.
    d. chromatid.
    e. chromatin strand.

Epithelial Tissue

As we continuc our study of the human body, we progress from the cellular level of organization to the tissue level. Tissues are masses of similar cells (and their extracellular *matrix*) that combine with other tissues to form *membranes* or *organs*. The systematic study of tissue types is called **histology.**

There are four basic tissue types in the human:

> epithelial tissue
> connective tissue
> muscle tissue
> nerve tissue

Each of these categories includes several subcategories, based on differences in structure and function. We will consider some of the major subcategories in this and the next several exercises. We will begin our exploration with a brief look at epithelial tissue types.

Before You Begin

☐ Read the appropriate chapter in your textbook.

☐ Set your learning goals. When you finish this exercise, you should be able to:
  o define the term *epithelium*
  o classify epithelial tissue types
  o identify and give examples of the major types of epithelia in figures and specimens

☐ Prepare your materials:
  o unlabeled photos of epithelial types (optional)
  o microscope
  o prepared microslides:
      simple squamous (surface)
      stratified squamous (nonkeratinized) c.s.
      stratified squamous (keratinized) c.s.
      simple cuboidal c.s.
      simple columnar c.s.
      pseudostratified (ciliated) c.s.
      transitional c.s.

☐ **Carefully** read the directions and safety tips for this exercise before starting any procedure.

A. How to Classify Epithelial Tissues

Epithelial tissue can have either of two basic roles: covering/lining and glandular. The first kind, covering/lining epithelium, is found in sheets that cover body structures (as in the outer layer of the skin) or line body spaces (as in the lining of the stomach). A characteristic of covering/lining epithelium is that it always has one side exposed, or free to face outward (covering) or inward (lining). Because the cells form a continuous sheet, they are held together very tightly and have very little matrix (extracellular material). The nonfree face of the epithelial sheet is attached to underlying connective tissue by the **basement membrane.** The basement membrane is a thin, gluelike layer that holds the epithelium in place while remaining highly permeable to water and other substances. This is important because epithelia do not have their own blood supply. Water and other important substances must diffuse between the epithelial cells and the underlying tissue through the basement membrane.

The second kind of epithelium is the type that forms glands. Glandular epithelium forms the functional portions of *exocrine glands* (glands that secrete substances into ducts that empty onto epithelial surfaces). It also forms *endocrine glands* (glands that secrete substances that diffuse into the bloodstream).

Epithelial tissue is usually categorized by its structure. The most common scheme classifies types by identifying the number of cell layers and the shape of the cells in the outer layer. We will practice classifying epithelia by using photomicrographs or drawings provided by your instructor, or the samples in figure 5.4.

☐ **1**   Randomly choose a sample from those provided.
☐ **2**   The first step in classifying an epithelial tissue is determining how many layers of cells constitute the sheet (fig. 5.1). The sample falls into one of three categories based on number of layers:

Free surface

Simple

Basement membrane

Connective tissue

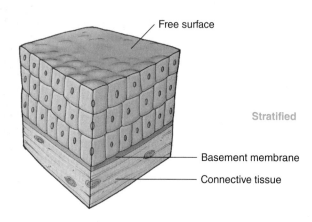

Free surface

Stratified

Basement membrane

Connective tissue

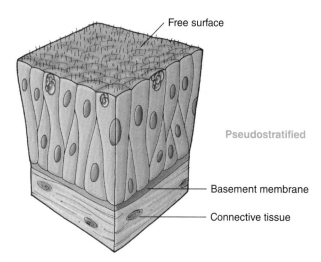

Free surface

Pseudostratified

Basement membrane

Connective tissue

Figure 5.1   Epithelial types are classified according to the number of cell layers.

o **Simple**—This category includes all tissues that have exactly one layer (or *stratum*) of cells.
o **Stratified**—This group includes epithelia with more than one layer of cells.
o **Pseudostratified**—The name of this category means "false stratified." This term is based on the fact that pseudostratified epithelium looks as if it were stratified but is really only one layer

deep. This illusion is caused by the way the cells seem to be pushed together, with some nuclei pushed into the upper portion of the cell and nuclei of other cells located in the bottom part. The nuclei then seem to form an upper and a lower row, giving the illusion of two rows of cells. Remember, cell membranes are often too thin for you to be able to see the boundary between epithelial cells.

□ **3** If you have determined that your sample is simple or stratified, skip to step 4. If your sample is pseudostratified, you must determine whether it is **ciliated** or **nonciliated.** Cilia are short, hairlike projections from each cell's free surface that propel material along the epithelial surface. You have now completely classified your sample and can try it again with a new sample.

□ **4** Look at the layer of cells along the sample's free surface. Determine which shape these cells have (fig. 5.2).
  o **Squamous**—Shaped like fish scales, these cells are much wider than they are tall when viewed in a cross section. Nuclei may be absent in cross sections.
  o **Cuboidal**—As their name implies, these cells are roughly cube-shaped, being about as tall as they are wide. Don't expect squared corners, as in a cube. Although they sometimes appear that way in a cross section, they more often resemble rounded squares or even circles.
  o **Columnar**—These columnlike cells are taller than they are wide when viewed in a cross section.
  o **Transitional**—The term *transitional* refers to the fact that these cells change shape as the need arises, being stretched or compressed into any of the three shapes already listed. Transitional cells are usually found as a hodgepodge of different cell shapes in membranes subject to a great degree of stretching and recoil. Transitional epithelia are always stratified.

If your sample is simple squamous, simple or stratified cuboidal, or stratified columnar, you are finished with this sample and can try another one. If you have simple columnar, move to step 5. If you have stratified squamous, skip to step 6.

□ **5** Determine whether the sample is ciliated or nonciliated. See step 3 for clarification. You are now finished classifying this sample and can try another sample.

□ **6** Determine whether your stratified squamous tissue is **keratinized** or **nonkeratinized.** In some stratified squamous tissues, the upper layers of cells are dead (evidenced by lack of nuclei) and filled with a tough, waterproof material called *keratin.* If your sample has

Free surface

Squamous

Connective tissue
Basement membrane

Free surface

Cuboidal

Connective tissue
Basement membrane

Free surface

Columnar

Connective tissue
Basement membrane

Figure 5.2 Examples of different cell shapes, here represented as simple epithelia.

a thick sheet of material with no visible nuclei along its free surface, it is probably keratinized. Nonkeratinized samples have nucleated squamous cells along the free edge.

B. Microscopic Specimens

In this activity, we will examine prepared slides of human epithelial tissues (fig. 5.3). Before beginning, review the microscopy techniques outlined in Lab Exercise 1.

SAFETY FIRST!

Avoid the hazards associated with frayed power cords and broken glass slides. Don't reach for anything while looking into the ocular, or you may knock over something (or someone).

For each of the epithelial tissues listed, examine at least one example in a prepared slide. Try to practice looking for the key characteristics by which each type can be classified. Sketch some representative examples in Lab Report 5.

Hint . . . In many (not all) prepared epithelia, the cytoplasm will appear either as a clear/cloudy area or a very pale pink. The cell membranes are usually faint but may be visible as medium pink lines. The nuclei are often stained dark pink to violet or bluish.

☐ **1** *Simple squamous epithelium*—As the name implies, simple squamous epithelium is a single layer of flattened cells. This type of epithelium forms the very thin lining found in the blood vessels, alveoli (air sacs) of the lungs, and other areas where thin membranes are required. Because it is so thin, simple squamous is well adapted for diffusion or filtration of water, gases, and other substances.

Landmark Characteristics

Viewed as a sheet from above, simple squamous epithelium looks like a two-dimensional layout of polygonal or rounded "tiles," each with a central nucleus (see Plate 2*b* in The Histology Minireference).

☐ **2** *Stratified squamous epithelium*—This tissue type is composed of multiple layers of cells: columnar cells along the basement membrane topped by cuboidal cells, then by squamous cells. Cells divide in the columnar layer and are pushed upward, where they are distorted into cuboidal, then squamous, cells. The cells on the surface slough off but are continually replaced by cells moving up from the bottom layer. Because of its thickness and its constant renewal, it is well adapted for protection. For example, stratified squamous epithelium is found in the outer part of the skin and the mucous linings of the mouth, vagina, and esophagus.

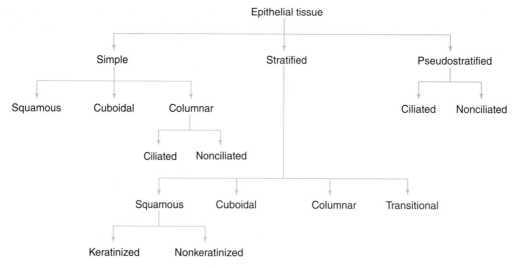

Figure 5.3   Classification scheme for epithelial tissues.

Landmark Characteristics

• *Nonkeratinized stratified squamous epithelium* typically has a dense concentration of nuclei (of columnar and cuboidal cells) near the basement membrane, becoming less dense toward the free surface. Most of the squamous cells near the free surface should have identifiable nuclei (see Plate 3*a* in The Histology Minireference).

• *Keratinized stratified squamous epithelium* will be from the epidermis (outer layer) of the skin. It will look similar to nonkeratinized specimens, except that it has a distinct additional layer overlying the top layer of squamous cells. This layer has no distinguishable nuclei (see Plate 7*a–b* in The Histology Minireference).

☐ **3** *Simple cuboidal epithelium*—Composed of a single layer of almost cubic cells, this type is found in secreting organs, such as glands. Simple cuboidal epithelium also forms the kidney tubules, where it is specialized for water reabsorption and ion movement.

Landmark Characteristics

The cuboid shape of cells is easily seen in cross sections of kidney tubules. Because the sample is formed by many tubules cut at an angle, you will see many circles and loops made of simple cuboidal epithelium. Similar specimens are seen in thyroid tissue and other glands and glandular ducts.

☐ **4** *Simple columnar epithelium*—Forming linings specialized for absorption and secretion, simple columnar epithelium is found in many parts of the body. For example, this type of epithelium lines portions of the reproductive tract, digestive tract, excretory ducts, and respiratory tract. A special cell that is often interspersed among the other columnar cells is the **goblet cell.** The goblet cell resembles its namesake, the wine goblet, in that it has a large, cuplike vesicle that may open onto the free surface. The cup has no wine but **mucus,** which the goblet cells produce and secrete in great quantity. Mucus has many functions, including the lubrication and protection of the epithelial lining.

Hint . . . The terms *mucous* and *mucus* are often confused. *Mucus* is a noun, naming a glycoprotein-water solution. *Mucous* is an adjective describing something covered with mucus. Thus, mucous membranes are covered with mucus.

Landmark Characteristics

Simple columnar sheets often line cavities with deeply folded or grooved walls. The specimen, then, will appear to zig-zag when viewed in a cross section. One surface of the sheet is always free, however, even though the free surfaces may fold back and touch one another. A goblet cell is easily recognizable by the very large bubble (vesicle) in the center or near the top. Because the vesicle contains clear, unstained mucus, it will appear more lightly colored than the surrounding material.

□ **5** *Ciliated pseudostratified epithelium*—This type of epithelium is a single layer of columnar cells that all attach to the basement membrane. However, some cells are short and don't quite reach the free surface. Nuclei of short cells are nearer the basement membrane than the nuclei of tall cells, so there appear to be two rows of cells. Like simple columnar epithelium, pseudostratified epithelium is found in both ciliated and nonciliated forms. It is found in many of the same general areas of the body as simple columnar epithelium: the upper throat, upper respiratory tract, and parts of the male urinary and reproductive tracts.

Landmark Characteristics

On initial examination, pseudostratified epithelium resembles simple columnar epithelium. However, the telltale double row of nuclei gives it away. As in simple columnar epithelium, goblet cells may be present. Cilia look like distinct, tiny hairs along the free surface. In some specimens, some of the cilia may have been damaged during preparation and so may have some areas of matted cilia or even "bald" spots. Under low magnification, the cilia look like "fuzz" on the free edge.

□ **6** *Transitional epithelium*—Transitional tissue is adapted for stretching, so it is found in areas subject to a great deal of elastic stress. You will most likely see specimens from the lining of the urinary bladder, which must stretch a great deal from time to time throughout the day.

Landmark Characteristics

Transitional epithelium resembles nonkeratinized stratified squamous epithelium at first glance. Transitional epithelium, however, often has rounded cuboidal cells in the top layer (rather than only squamous cells). Depending on the individual specimen, you may see a great variety of shapes scattered throughout the tissue. This gives transitional tissue a rather unorganized appearance, compared with the other epithelial types. Occasionally, transitional cells have two nuclei—an unusual characteristic that can help you distinguish transitional from other epithelial types that have just one nucleus per cell. Review epithelial tissue types by filling in figure 5.4.

Clinical Application

Researchers are always looking for new ways to deliver therapeutic drugs to the bloodstream for distribution throughout the body. Currently, biomedical research teams are using epithelial tissue cultures to investigate a relatively new approach to drug introduction. For some time, science has known that certain compounds are easily absorbed by epithelial membranes and picked up by the blood. For example, patches of material containing drugs that inhibit motion sickness have been used by tourists on ocean cruises. Patients with heart disease or burns often receive their drugs by means of similar patches. Some women receive birth control drugs using patches applied to the skin. Today, however, researchers are looking for wider and more varied uses of epithelial absorption as a drug-delivery method.

Based on what you know of the different epithelial tissue types, which are the best candidates for this type of drug delivery? Explain your choices.

How could a clinician actually apply epithelial absorption techniques? Try to think of strategies other than the skin-patch approach.

Explain some of the harmful side effects of smoking tobacco, cocaine, and marijuana in terms of epithelial absorption.

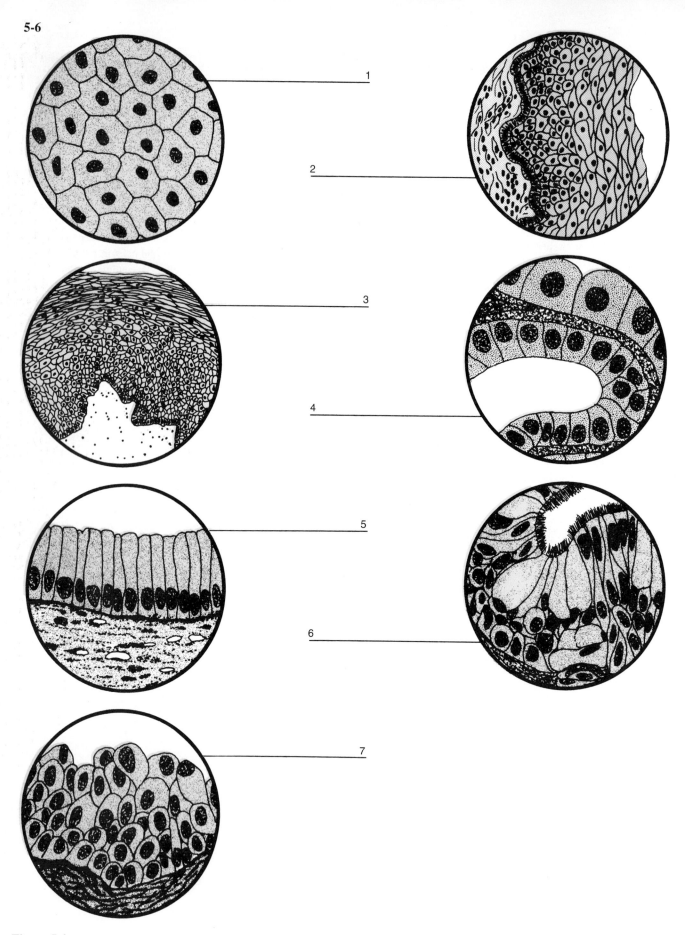

Figure 5.4 Identify each of the tissue types represented by these figures.

LAB REPORT 5

Epithelial Tissue

Name _____

Date _____   Section _____

Activity A

Trial Number	Identifying Code*	Tissue Type	Notes
1			
2			
3			
4			
5			
6			
7			

*The identifying code is the code number (letter) assigned to your sample by the instructor. If you are using figures from this or another book, put the plate and/or figure number in this column.

Specimen: *simple squamous epithelium (surface view)*
    Total magnification: _____

Specimen: *stratified squamous epithelium (keratinized)*
    Total magnification: _____

Specimen: *stratified squamous epithelium (nonkeratinized)*
    Total magnification: _____

Specimen: *simple cuboidal epithelium*
    Total magnification: _____

Specimen: *simple columnar epithelium*
   Total magnification: _____

Specimen: *transitional epithelium*
   Total magnification: _____

Specimen: *ciliated pseudostratified epithelium*
   Total magnification: _____

Specimen: _____
   Total magnification: _____

Figure 5.4 *(from p. 32)*

_____ 1

_____ 2

_____ 3

_____ 4

_____ 5

_____ 6

_____ 7

Identify

_____ 1

_____ 2

_____ 3

_____ 4

_____ 5

_____ 6

_____ 7

_____ 8

Identify

Identify the term associated with each description.

1. This type of cell is found in simple columnar and pseudostratified tissue that secretes mucus.
2. One of two kinds of glands, it secretes substances onto epithelial surfaces.
3. A cellular organelle, it is a projection that moves substances along the surface of a cell.
4. This type of epithelial tissue is especially well adapted to excessive stretching.
5. Mucous membranes secrete this water-based protein solution.
6. This tough, waterproof material is found in the upper layers of some examples of stratified squamous epithelium.
7. This type of epithelial tissue is likely to form glands.
8. This type of epithelial tissue is likely to be found on the palm of the hand.

LAB EXERCISE 6

Connective Tissue

Some tissues are called **connective tissues** because they act as connections among various other tissues. Bone, cartilage, and fibrous connective tissues actually hold parts together or support them in some way. Blood tissue connects other tissues in the sense of transporting materials between them.

An additional function of some connective tissues is storage. For example, bone tissue is an important storage site for calcium and phosphorus, minerals vital to proper function in many parts of the body.

In this exercise, we will learn to identify some of the more common tissue types found in the human body.

Before You Begin

☐ Read the appropriate chapter in your textbook.

☐ Set your learning goals. When you finish this exercise, you should be able to:
  o describe the general pattern of structure in connective tissues
  o classify major connective tissue types
  o identify examples of connective tissue types in figures and specimens

☐ Prepare your materials:
  o microscope
  o prepared microslides:
      dense fibrous (regular)
      dense fibrous (irregular)
      loose fibrous (areolar)
      adipose
      hyaline cartilage
      fibrocartilage
      elastic cartilage
      compact bone
      cancellous bone
      blood smear

  o fresh animal connective tissue samples (as available)

☐ **Carefully** read the directions and safety tips for this exercise before starting any procedure.

A. Classifying Connective Tissues

A structural feature common to all connective tissues is the dominance of the matrix, or extracellular material. Recall that epithelial tissues have almost no extracellular material separating individual cells. Connective tissue cells are often widely separated by one of three basic types of matrix:

☐ **1** *Protein fiber matrix* is extracellular material composed of many substances but with a dominance of protein fibers. **Collagen** is a common protein, forming bundles of tough, flexible fibers. Because they have a whitish color, collagen fibers are often called *white fibers.* **Elastin,** a stretchy, fibrous protein, forms thick, single fibers in connective tissue matrices. Elastin fibers are sometimes called *yellow fibers.* Identify these categories of protein fiber-matrix connective tissues in figure 6.1:
  o **Fibrous**—Fibrous tissues are categorized as either **dense fibrous** or **loose fibrous,** depending on the density of protein fibers in the matrix. Dense fibrous tissue can be classified as **dense collagenous fibrous** or as **dense elastic fibrous,** depending on whether they have mostly collagen fibers or mostly elastic fibers in their matrix. Dense fibrous tissue can also either be **regular dense fibrous** (having regular parallel bundles of fibers) or **irregular dense fibrous** (having an irregular hodgepodge of fibers). Loose fibrous tissue is also called **areolar tissue.**
  o **Adipose**—Adipose tissue is often simply called *fat tissue* because its primary function is the storage of fat (for later use or for body support). Adipose tissue is actually a modified form of areolar tissue, with fat storage cells having been filled with stored lipids and expanded into the extracellular spaces.
☐ **2** *Protein/ground substance matrix* is extracellular material that has some protein fibers in it but also a great deal of nonfibrous protein and other substances. Identify these categories of protein/ground substance tissues in figure 6.1:
  o **Cartilage**—The matrix of cartilage is a combination of fibers and ground substance that

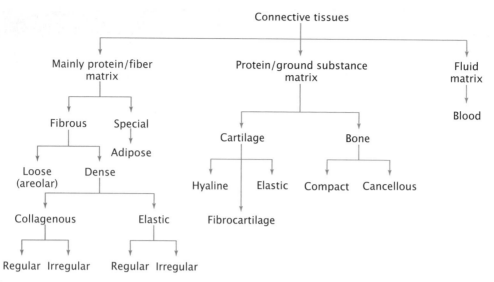

Figure 6.1   A classification scheme for connective tissues.

gives it a rubbery quality. **Hyaline cartilage** has a moderate amount of collagen fiber in its matrix. **Fibrocartilage,** as its name implies, has a large amount of collagen in its matrix. **Elastic cartilage** is distinguished by the presence of elastin fibers, giving it a stretchy quality.

○ **Bone**—There are two broad categories of bone: **compact bone** and **cancellous (spongy) bone.** Both have a matrix of collagen fibers encrusted with mineral crystals that give it a solid consistency. Compact bone forms rather large, dense pieces of bone matrix. Cancellous bone forms thin, narrow beams of hard bone matrix in which **red bone marrow** can be supported.

☐ **3**   *Fluid matrix* is composed of a water-based solution with a fluid consistency. **Blood** is the major type of fluid matrix connective tissue. Blood cells are suspended within the fluid *plasma* and can slide past one another freely. Identify fluid-matrix tissue in the general scheme presented in figure 6.1.

Hint . . .   Many cells found in connective tissues are named according to their action. Cells that produce matrix often have the suffix *-blast* ("make"). Cells that destroy matrix during remodeling have the suffix *-clast* ("break"). Cells that are in a relatively inactive mode have a *-cyte* ("cell") suffix. The first part of the name tells the specific kind of matrix involved: *fibro-* ("fiber"), *chondro-* ("cartilage"), or *osteo-* ("bone"). Thus, *fibroblasts* make fibers, *chondroblasts* manufacture cartilage matrix, and *osteoblasts* lay down bone matrix.

Histologists classify tissues for convenience in naming them and for better identification of their nature. However,

different histologists use slightly different schemes, depending on the context in which they are working. The histology schemes used in this manual are fairly universal.

B. Microscopic Specimens

SAFETY FIRST!

Avoid electrical hazards while using the microscope. Be sure to exercise care in dealing with broken glass slides. Don't reach for slides or other objects while you are looking into the ocular, or you may knock over something (or someone).

This activity asks you to observe representative connective tissue types in prepared specimens. For each type listed in the following steps, locate an example in a slide and sketch it in Lab Report 6. Also, identify the connective tissues in figures 6.2 and 6.3.

☐ **1**   *Dense fibrous connective tissue*—As its name indicates, this tissue is a dense arrangement of fibers. They may be collagen fibers or elastic fibers. Thus, there may be *dense collagenous* or *dense elastic* connective tissue. Their arrangement may be regular (roughly parallel) or irregular (swirling or random). Interspersed among the fibers of a mature tissue are **fibrocytes.** Irregular dense fibrous tissue forms the lower layer of the skin (dermis), much of the body's fascia, and the capsules of many organs. Regular dense fibrous tissue is used for structures that require a better-engineered connection between parts that are pulled with great force. For example, this tissue forms *tendons* (connecting muscle to bone) and *ligaments* (connecting bone to bone).

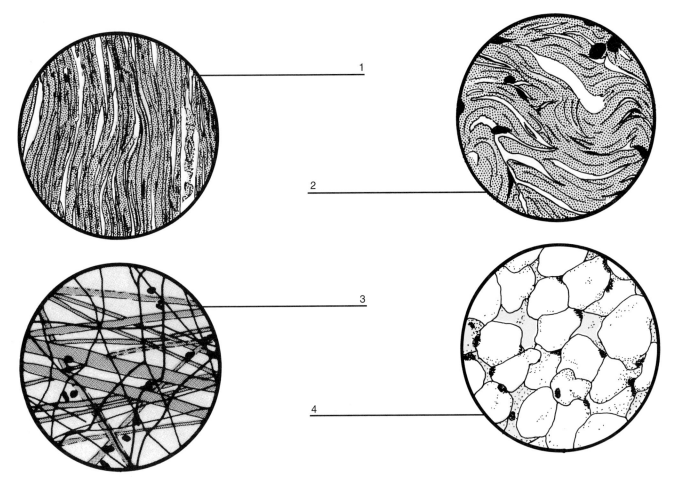

Figure 6.2 Identify the connective tissues represented in these drawings.

through it. A variety of cell types may be found here: fibrocytes, adipose cells, and white blood cells.

Landmark Characteristics

- Regular dense fibrous tissue usually has a roughly parallel arrangement of either collagen or elastic fibers. Collagen appears in wavy bundles stained pink or bluish. Elastin appears as thick fibers stained dark violet or blue. Darkly stained fibrocytes are scattered between fibers or fiber bundles, often in groups (see Plate 5c in The Histology Minireference).

- Irregular dense fibrous tissue is virtually the same as regular tissue in composition. Irregular tissue is different because the fibers appear as chaotic swirls rather than parallel lines (see Plate 5b–c in The Histology Minireference).

Landmark Characteristics

Areolar tissue's widely spaced fibers and variety of cell types make it an easy tissue to identify. The elastin fibers appear as dark, thick, jagged strands. Collagen fibers form bundles that often appear as hazy pink or light purple criss-crossing lines (see Plate 5a in The Histology Minireference).

☐ **2** *Loose fibrous (areolar) connective tissue*—Areolar (meaning "spacious") tissue forms loose bonds between other tissues. For example, under the skin, it allows the skin to be slid over, or pulled from, the underlying muscle to some degree. Both collagen and elastin fibers are found in this tissue but are widely spaced. Blood vessels, as well as nerves, course

☐ **3** *Adipose tissue*—Adipose cells are specialized to store lipids in large vesicles. The vesicle can be so large that it pushes the nucleus and other organelles up to the cell membrane, which enlarges to accommodate the large cell volume. Some histologists classify adipose tissue as a subcategory of areolar tissue. If the adipose cells in areolar tissue enlarge (due to increased storage of fat), they crowd the fibers and other cells, eventually forming adipose tissue. Thus, adipose tissue has very little matrix, compared with other connective tissue types.

1

2

3

4

5

6

Figure 6.3 Identify these connective tissues.

Adipose tissue is found wherever areolar tissue is found but is most often seen around the heart and under the skin. It not only stores lipids for later use but also serves as support (as in the breasts), as insulation (under the skin), and as a cushion (you're sitting on it).

Landmark Characteristics

Unlike other connective tissue cells, adipose cells are very close to one another. The large fat globule inside each cell pushes the cytoplasm and organelles into a thin, dark ring around the inside of the cell membrane. The nucleus often appears as a bulge in the ring. The most obvious characteristic of this tissue is the presence of large fat vesicles, which generally look clear, yellowish, or light pink. The overall appearance is that of a host of large bubbles (see Plate 5d in The Histology Minireference).

☐ **4** *Hyaline cartilage*—All cartilage is distinctive in the consistency of its semisolid, rubbery matrix of protein fibers mixed with other substances. **Chondrocytes** are scattered throughout the matrix in little pockets called **lacunae** (meaning "lakes"). Hyaline cartilage has a moderate amount of collagen, giving it a great deal of toughness along with its cushiony quality. This cartilage type forms the bulk of the fetal skeleton (before it is replaced by bone) and continues to be the most abundant type of cartilage throughout life. It forms the thin, rubbery layer over the ends of long bones and is found in parts of the larynx, nose, and trachea.

Landmark Characteristics

The collagen fibers of hyaline cartilage are not distinct in the matrix, which has a smooth, pinkish or lavender appearance in many preparations. The chondrocytes are usually pink to violet and appear to have shriveled within their respective lacunae. Because of chondrocyte shrinkage, a clear ring appears around the inside of many lacunae (see Plate 6b in The Histology Minireference).

☐ **5** *Fibrocartilage*—The name of this tissue indicates its high concentration of collagen fibers. These fibers give the tissue a distinctive fibrous appearance. Fibrocartilage has a more rigid, less rubbery consistency than other cartilage types. Fibrocartilage forms the disks between vertebrae and may be found at other semimovable joints.

Landmark Characteristics

Fibrocartilage may appear alongside hyaline cartilage and sometimes looks very much like it. However, the distinct fibrous appearance of fibrocartilage's matrix is the determining factor (see Plate 6c in The Histology Minireference).

☐ **6** *Elastic cartilage*—As its name implies, elastic cartilage has a large proportion of elastin fibers in its matrix. Elastic cartilage is found in structures where springiness is desirable in the support material. For example, the elasticity of the pinna (ear flap) is provided by elastic cartilage.

Landmark Characteristics

In many preparations, elastin fibers stain very darkly. This makes their presence easy to detect and allows one to distinguish elastic cartilage without any problem (see Plate 6d in The Histology Minireference).

☐ **7** *Compact bone*—Compact bone is formed by solid, cylindrical units, called **osteons,** packed tightly together. The osteon, or **haversian system,** consists of multiple concentric layers of hard bone matrix, with cells sandwiched between layers. Each layer is a **lamella** (pl. **lamellae**). **Osteocytes** are literally trapped within lacunae between the lamellae. The osteocytes were once active **osteoblasts** but have trapped themselves in the solid matrix they formed. The lamellae are centered around the blood vessels of the **central canal** or **haversian canal.** The cells transport materials to and from the canal by way of tiny **canaliculi** ("small canals") that connect the osteocytes to each other and to the canal.

Landmark Characteristics

A cross section of compact bone has rings of lamellae surrounding several adjacent *central* canals. The lamellae resemble rings in an onion slice. The central canals are either clear or nearly black, the lamellae buff to orange, and the osteocytes brown or black. The canaliculi often appear as wavy hairlines radiating from the lacunae (see Plates 8b and 10a in The Histology Minireference).

□ **8** *Cancellous bone*—Cancellous bone is easily identified by its open, latticelike structure. Thin plates of bone matrix, with a scattering of osteocytes trapped within lacunae, form structural beams that have great strength despite the open spaces. These beams of hard bone are called **trabeculae.** Because cancellous bone has open spaces, it is sometimes called *spongy bone.* This name can be misleading because one might think spongy bone is as soft as a bath sponge. It isn't soft at all because it has hard trabeculae. The spaces are filled with *myeloid tissue,* a type of blood tissue that produces new blood cells. Myeloid tissue is called *red bone marrow.*

Landmark Characteristics

Cancellous bone is distinguished by its rather disorganized array of trabecular beams of bone surrounded by myeloid tissue. The bone pieces may look like slivers of compact bone, with lamellae that often don't form complete circles. The myeloid tissue is a scattering of blood cells, which appear as tiny, dark circles. Myeloid tissue may also have a netlike formation of very thin collagen fibers called *reticular fibers.* In some preparations, the bone tissue is pink, and the myeloid cells are dark red (see Plates 8*a* and 10*b* in The Histology Minireference).

□ **9** *Blood*—Blood tissue is a fluid-matrix connective tissue characterized by a variety of cell types. Tiny **red blood cells (RBCs),** tinier fragments called **platelets,** and larger **white blood cells (WBCs)** may all be seen in a blood-smear preparation. Of course, blood is found within the circulation vessels and has many functions. Blood transports and exchanges materials, serves in immune protection of the body, and helps regulate body temperature, among other functions. We will study blood in more detail in Lab Exercise 25.

Landmark Characteristics

A prepared blood smear is a drop of blood smeared on a slide and stained. The red blood cells are the more numerous, smaller cells stained pink to orange-red. RBCs have no nuclei. The white blood cells are much larger, with distinct, often lobed or segmented nuclei. Lab Exercise 25 has more complete details (see Plate 15*a–d* in The Histology Minireference).

C. Gross Tissue Examples

Gross (large) examples of connective tissues often give the learner a more complete mental image of connective tissue types. If gross specimens are available, explore the texture, consistency, and other physical characteristics of the connective tissues listed.

SAFETY FIRST!

Any animal tissue at room temperature can harbor growing colonies of microorganisms. As a safety precaution, *treat the samples as if they were contaminated* with harmful microbes. Use gloves or dissection tools when handling the samples, and disinfect any surfaces they touch. Use protective eyewear.

Type	Physical Characteristics
Dense fibrous (regular) tissue	
Dense fibrous (irregular) tissue	
Loose fibrous (areolar) tissue	
Adipose tissue	
Hyaline cartilage	
Fibrocartilage	
Elastic cartilage	
Compact bone	
Cancellous bone	

Hint . . . Learning tissue identification is not as hard as it may first appear. The trick is to practice *frequently.* Besides practicing in the lab, try using flash cards made with your sketches or photos.

Connective Tissue

Name _____

Date _____ Section _____

Specimen: *dense fibrous (regular) connective tissue*
　　　Total magnification: _____

Specimen: *adipose tissue*
　　　Total magnification: _____

Specimen: *dense fibrous (irregular) connective tissue*
　　　Total magnification: _____

Specimen: *hyaline cartilage*
　　　Total magnification: _____

Specimen: *loose fibrous (areolar) tissue*
　　　Total magnification: _____

Specimen: *fibrocartilage*
　　　Total magnification: _____

Specimen: *elastic cartilage*
Total magnification: _____

Specimen: *cancellous bone*
Total magnification: _____

Specimen: *compact bone*
Total magnification: _____

Specimen: *blood smear*
Total magnification: _____

Figure 6.2 *(from p. 37)*

_____ 1

_____ 2

_____ 3

_____ 4

Figure 6.3 *(from p. 38)*

_____ 1

_____ 2

_____ 3

_____ 4

_____ 5

_____ 6

Matching

_____ 1   _____ 5   _____ 9

_____ 2   _____ 6   _____ 10

_____ 3   _____ 7

_____ 4   _____ 8

Matching

Answer may be used more than once or not at all.

a. dense fibrous connective tissue
b. adipose tissue
c. hyaline cartilage
d. fibrocartilage
e. elastic cartilage
f. compact bone
g. cancellous bone

1. The cartilage type with a great deal of collagen in the matrix
2. The tissue type that is actually modified areolar tissue
3. The tissue type that stores lipid molecules
4. The tissue type composed of haversian systems
5. The most common type of cartilage
6. Strong tissue that forms tendons and ligaments
7. The tissue type associated with red bone marrow
8. The tissue type that forms strong membranes
9. Connective tissue that forms the disks between vertebrae
10. The tissue type that forms hard mineral trabeculae

LAB EXERCISE 7

Muscle and Nerve Tissue

Our exploration of the four basic human tissue types concludes with this study of muscle and nerve tissue. Detailed study of these tissue types is more appropriate for a later part of this course, but it is essential to achieve a basic understanding of these tissues now.

Before You Begin

☐ Read the appropriate chapter in your textbook.

☐ Set your learning goals. When you finish this exercise, you should be able to:
  o distinguish among the three basic types of muscle tissue
  o identify the primary components of nerve tissue
  o recognize muscle and nerve tissue types in figures and specimens

☐ Prepare your materials:
  o microscope
  o prepared microslides:
      skeletal muscle l.s. (longitudinal section)
      cardiac muscle l.s.
      smooth muscle l.s.
      spinal cord smear

☐ **Carefully** read the directions and safety tips for this exercise before starting any procedure.

A. Muscle Tissue

All three muscle tissue types contain cells with the ability to contract. Contraction in muscle tissue allows a muscle organ to pull with great force. Such pulling can move the skeleton or squeeze the contents of a hollow organ. In addition to this common function, all muscle tissues are composed of long, cylindrical cells called **muscle fibers.** Muscle fibers are living cells, unlike connective tissue fibers, which are nonliving protein fibers (fig. 7.1).

SAFETY FIRST!

Observe the usual precautions when using the microscope and slides.

☐ 1 **Skeletal muscle** gets its name from the fact that it forms muscular organs that attach to the skeleton and move its parts. This tissue is also called *striated voluntary muscle* because it has striations (stripes) and can be controlled by the conscious mind. Skeletal muscle cells are large, fiberlike cells with fine cross-stripes and many nuclei per cell. Skeletal muscle fibers and bundles of muscle fibers often have a coating of fibrous connective tissue.

Landmark Characteristics

Skeletal muscle preparations include very large cells with the characteristic striped pattern. The striping is very fine, so good focusing technique is critical with this specimen. The muscle fibers are generally parallel to one another in a longitudinal section (l.s.), a cut made lengthwise, with the stripes at right angles across each fiber. Unlike the other types of muscle, skeletal muscle fibers have many nuclei per fiber. The nuclei are often up against the fiber membrane. Some fibrous connective tissue may be seen between the muscle fiber (see Plate 11*a–b* in The Histology Minireference).

☐ 2 **Cardiac muscle** is also known as *striated involuntary muscle.* This tissue also has striations, though less distinct than in skeletal muscle. Cardiac muscle is involuntary in the sense that subconscious mechanisms regulate its contraction. Cardiac muscle is found only in the walls of the heart. Because it must encircle and compress the heart chambers with great strength to pump blood, cardiac muscle requires some features not found in other muscle types. For example, the individual fibers are branched, allowing the fibers to mesh with other cells at different layers. Also, cardiac muscle fibers are fused end-to-end by **intercalated disks.** The branching and fusing gives a group of cells the ability to functionally imitate a giant cell encircling one or more chambers of the heart.

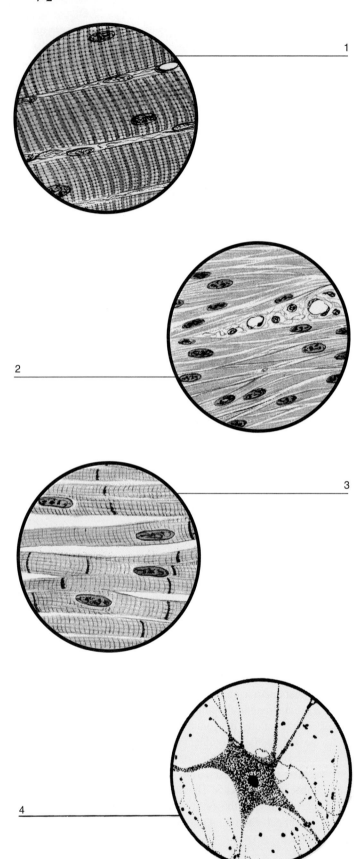

Landmark Characteristics

Cardiac muscle striations are less distinct than those in skeletal muscle. Cardiac fibers have single nuclei, usually have branches, and do not taper at their ends. Many fibers attach end-to-end via intercalated disks, which appear as fine, dark (sometimes purple) lines at a right angle to a seemingly continuous fiber (see Plate 11c in The Histology Minireference).

☐ **3** **Smooth muscle** gets its name from the fact that it has no distinct striations. Like cardiac muscle, it also is an involuntary muscle type. Smooth muscle is found in the walls of hollow organs, such as digestive organs and blood vessels. This tissue type is composed of long, threadlike cells, each with a single nucleus. The cells are generally parallel with one another and with the edge of the wall in which they are embedded.

Landmark Characteristics

Unlike other muscle cells, smooth muscle fibers are unstriated and have single nuclei. In some preparations, the cells are pink and the nuclei purple or black. At first glance, you may confuse smooth muscle tissue with dense fibrous (regular) connective tissue. Smooth muscle is not wavy, as dense fibrous tissue sometimes is, and has a more even distribution of nuclei than dense fibrous tissue (see Plate 11d in The Histology Minireference).

B. Nerve Tissue

Nerve tissue composes organs of the nervous system: the brain, spinal cord, and nerves. Two basic types of cells are found in this tissue: **neurons** (impulse-conducting cells) and **neuroglia** (support cells). Neurons are large cells with nucleated *bodies* and projections called *axons* and *dendrites*. There are many types of neuroglia, but they are generally smaller than neurons, which they outnumber by several times. Neuroglia surround and support neurons physically or biochemically.

Landmark Characteristics

In a smear of spinal cord tissue, the neurons and neuroglia are randomly scattered on the slide. The neurons are extremely large, each with a body having a single nucleus. Neuron projections crisscross throughout. Neuroglia appear as tiny, dark dots (see Plate 16b in The Histology Minireference).

Figure 7.1 Identify these tissue types by labeling them.

LAB REPORT 7

Muscle and Nerve Tissue

Name _____

Date _____   Section _____

Specimen: *skeletal muscle*   Total magnification: _____

Specimen: *smooth muscle*   Total magnification: _____

Specimen: *cardiac muscle*   Total magnification: _____

Specimen: *spinal cord smear*   Total magnification: _____

Figure 7.1 *(from p. 44)*

_____ 1

_____ 2

_____ 3

_____ 4

Multiple choice

_____ 1

_____ 2

_____ 3

_____ 4

_____ 5

_____ 6

_____ 7

_____ 8

Multiple choice

1. A voluntary type of muscle tissue is
    - a. skeletal.
    - b. cardiac.
    - c. smooth.
    - d. *a* and *b*.
    - e. *a* and *c*.

2. An involuntary muscle tissue is
    - a. skeletal.
    - b. cardiac.
    - c. smooth.
    - d. *b* and *c*.
    - e. *a* and *c*.

3. Which of these muscle tissues is striated?
    - a. skeletal
    - b. cardiac
    - c. smooth
    - d. *a* and *b*
    - e. all

4. Which of these is most likely to be found in the wall of the urinary tract?
    - a. skeletal
    - b. cardiac
    - c. smooth

5. Cardiac muscle fibers are connected end-to-end by means of
    - a. Velcro.
    - b. intercalated disks.
    - c. striations.
    - d. branching.

6. Which of these cells is likely to have multiple nuclei?
    - a. skeletal muscle
    - b. cardiac muscle
    - c. smooth muscle
    - d. neuron
    - e. *a* and *c*
    - f. *a, c,* and *d*

7. Nerve tissue contains cells called
    - a. fibers.
    - b. neurons.
    - c. neuroglia.
    - d. striated cells.
    - e. *b* and *c*.

8. Nerve tissue forms the bulk of the
    - a. brain.
    - b. heart.
    - c. digestive tract.
    - d. vertebrae.
    - e. *a* and *d*.

Organization of the Body

Many explorers use maps, figures, and photographs to help orient themselves to the terrain to be explored. The anatomist uses maps, figures, and photos to explore the body and its parts. The use of maps and other aids to find a geographical position is called *orienteering* and is a useful analogy to human anatomical study. At the beginning of this exercise, you will learn how anatomical "maps" and models are read. Later in this exercise, you will become familiar with the major body systems and some of their organs so that you will be comfortable with the "lay of the land" in the human body.

Before You Begin

□ Read the appropriate chapter in your textbook.

□ Set your learning goals. When you finish this exercise, you should be able to:
  o use anatomical terms correctly
  o appreciate the nature of an anatomical section
  o understand the basic plan of the human body
  o identify the major body cavities
  o list the major systems of the body, their principal organs, and their primary functions

□ Prepare your materials:
  o dissectible human torso model (or comparable charts)
  o models or figures showing different anatomical sections

□ **Carefully** read the directions and safety tips for this exercise before starting any procedure.

A. Planes and Sections

All terms describing the anatomy of organisms assume that the body is in the classic **anatomical position.** For the human, that means standing, facing the viewer. The hands are held down along the side of the trunk, with the palms facing forward.

It is often useful to show a figure of a sectioned human body or organ. A **section** refers to a part cut in a **plane.** A plane is a geometrical concept referring to an imagined flat surface. The terms *transverse section* and *cross section (c.s.),* for example, refer to a part cut cross-wise. An *oblique section* is

like a *transverse section* but cut at a diagonal angle. A *longitudinal section (l.s.)* is a cut made lengthwise (fig. 8.1). These terms are useful only in limited circumstances, since they don't really identify whether the cuts are made top to bottom, front to back, or side to side.

There are three anatomical planes used to describe sections of the body:

□ **1 Sagittal plane**—A sagittal plane extends from front to back and top to bottom, dividing the body into left and right portions. A *median plane,* or *midsagittal plane,* refers to a sagittal plane that divides the body into exactly equal left and right portions.

□ **2 Frontal plane**—This plane, also called a **coronal plane,** divides the body into front and back portions.

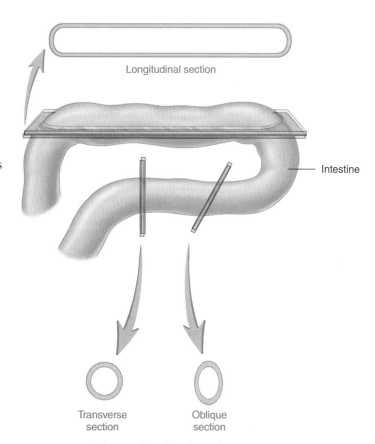

Figure 8.1 Planes of section through an organ.

TABLE 8.1   *Directional Terms*

Term	Definition	Example of Usage
Left	To the left of the body (not *your* left, the subject's)	The stomach is to the *left* of the liver.
Right	To the right of the body or structure being studied	The *right* kidney is damaged.
Lateral	Toward the side; away from the midsagittal plane	The eyes are *lateral* to the nose.
Medial	Toward the midsagittal plane; away from the side	The eyes are *medial* to the ears.
Anterior	Toward the front of the body	The nose is on the *anterior* of the head.
Posterior	Toward the back (rear)	The heel is *posterior* to the toes.
Superior	Toward the top of the body	The shoulders are *superior* to the hips.
Inferior	Toward the bottom of the body	The stomach is *inferior* to the heart.
Dorsal	Along (or toward) the vertebral surface of the body	Her scar is along the *dorsal* surface.
Ventral	Along (or toward) the belly surface of the body	The navel is on the *ventral* surface.
Caudad (caudal)	Toward the tail (used in animals only)	The neck is *caudad* to the skull.
Cephalad	Toward the head (used in animals only)	The neck is *cephalad* to the tail.
Proximal	Toward the trunk (describes relative position in a limb or another appendage)	This joint is *proximal* to the toenail.
Distal	Away from the trunk or point of attachment	The hand is *distal* to the elbow.
Visceral	Toward an internal organ; away from the outer wall (describes positions inside a body cavity)	This organ is covered with the *visceral* layer of the membrane.
Parietal	Toward the wall; away from internal structures	The abdominal cavity is lined with the *parietal* peritoneal membrane.
Deep	Toward the inside of a part; away from the surface	The thigh muscles are *deep* to the skin.
Superficial	Toward the surface of a part; away from the inside	The skin is a *superficial* organ.
Medullary	Refers to an inner region, or *medulla*	The *medullary* portion of the organ contains nerve tissue.
Cortical	Refers to an outer region, or *cortex*	The *cortical* area produces hormones.

□ **3** **Horizontal plane**—Also called a **transverse plane,** the horizontal plane divides the body into top and bottom portions.

Using the models or figures provided, find at least three examples of each of the sections described. For each, ask yourself what perspective the section gives that a section cut along a different plane does not give.

Hint . . . The anatomical position of a four-legged animal, such as a rat or fetal pig, is standing on all four limbs, head facing forward (see Lab Exercises 41 and 42 for examples).

B. Anatomical Directions

To locate structures within a body, you must use *directional terms*. Actually, you use these kinds of terms all the time: left, right, up, down, north, south, for example.

□ **1** Review the directional terms given in table 8.1. Notice that they are grouped in relative pairs. Each member of a pair is the opposite, or complement, of the other member of the pair. For example, *right* is the opposite direction of *left*.

To make the reading of anatomical figures a little easier, an *anatomical compass* is used throughout this book. On many figures, you will notice a small compass rosette similar to those on geographical maps. Rather than being labeled N, S, E, and W, the anatomical rosette is labeled with abbreviated anatomical directions.

☐ **2** Review this list of directional terms and abbreviations:

> **A** = Anterior
> **I** = Inferior
> **L** (opposite **R**) = Left
> **L** (opposite **M**) = Lateral
> **M** = Medial
> **P** = Posterior
> **R** = Right
> **S** = Superior

☐ **3** Review the examples of how the anatomical compass is used in illustrations by looking at figure 8.2a–d. Test your knowledge of directions and the use of the anatomical compass by labeling the rosettes given in figure 8.2e–j.

Hint . . . A handy summary of anatomical terms and their usage is found on the inside front cover of this lab manual, for easy location later in your studies.

C. Body Cavities and Regions

The inside of the human body contains the **viscera,** or internal organs (fig. 8.3). The viscera are found in any of a number of **cavities** (spaces) within the body. Because some cavities are so large, they are subdivided into smaller units.

☐ **1** Using a dissectible torso model, find these cavities of the body (and organs within):
- **cranial cavity**—within the skull
  organ: *brain*
- **spinal cavity**—within the vertebral column
  organ: *spinal cord*

☐ **2** Using the torso model, find these divisions and organs of the cavities of the trunk:
- **thoracic cavity**—within the rib cage
  - **pleural cavities**—left one-third and right one-third of the thoracic cavity
    organ: *lung*
  - **mediastinum**—middle one-third of the thorax
    organs: *heart, trachea, esophagus*

- **abdominopelvic cavity**—from the diaphragm to the bottom of the trunk (fig. 8.4)
  - **abdominal cavity**—from the diaphragm to the rim of the pelvic bones
    organs: *stomach, liver, most of the intestines, pancreas, spleen, kidneys*
  - **pelvic cavity**—from the pelvic rim to the floor of the trunk
    organs: *portions of the intestines, ovaries, uterus, urinary bladder*

☐ **3** Because the abdominopelvic cavity is so large, and contains so many organs, it is often convenient to subdivide it into nine **abdominopelvic regions.** The regions are bounded by a grid made by imagining two horizontal planes (one just below the ribs, the other just above the hip bones) and two sagittal planes (each just medial to a nipple). This arrangement forms a three-dimensional, tic-tac-toe grid in the abdominopelvic cavity. Identify the approximate locations of each of the nine regions on a model of the human torso.
- **right hypochondriac region**—top right region (*hypochondriac* means "below [rib] cartilage")
- **epigastric region**—top middle region (*epigastric* means "near the stomach")
- **left hypochondriac region**—top left region
- **right lumbar region**—middle right region (*lumbar* refers to lumbar vertebrae in lower back)
- **umbilical region**—central region (*umbilical* refers to the umbilicus, or navel)
- **left lumbar region**—middle left region
- **right iliac region**—lower right region (*iliac* refers to ilium, the bowl-like part of the hip bone)
- **hypogastric region**—lower middle region (*hypogastric* means "below the stomach")
- **left iliac region**—lower left region

D. Surface Regions

There are hundreds of terms that describe specific locations on the surface of the human body. These names are useful for identifying not only surface features but also underlying muscles, bones, nerves, and blood vessels (fig. 8.5). In this activity, locate regions named by a few of the more common terms.

☐ **1** Locate the following surface regions on the anterior aspect of a human model or figure:
- **abdominal**—area overlying the abdominal cavity
- **axillary**—armpit
- **brachial**—upper arm

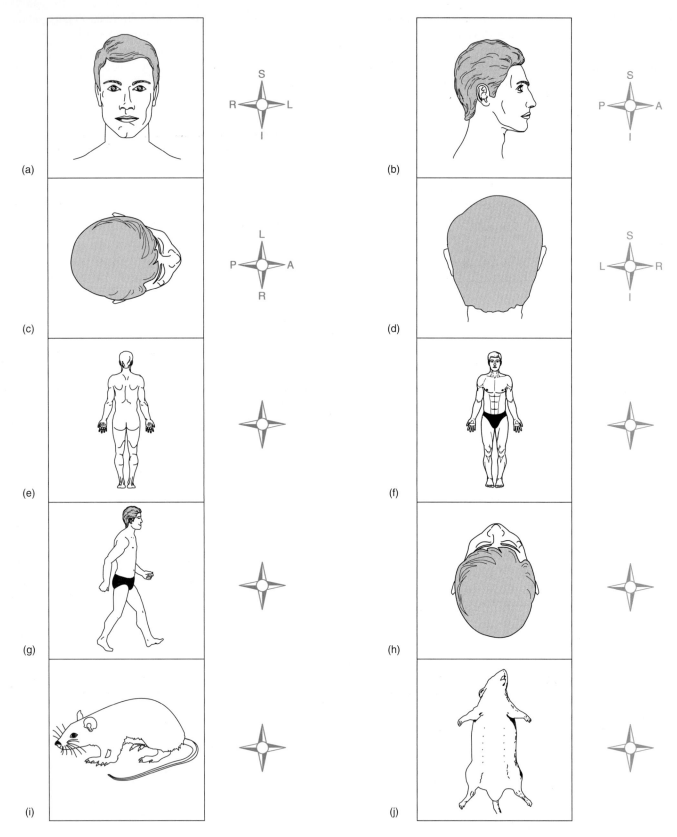

Figure 8.2 Look at the examples of labeling anatomical directions (*a* through *d*); then label the rosettes in *e* through *j*.

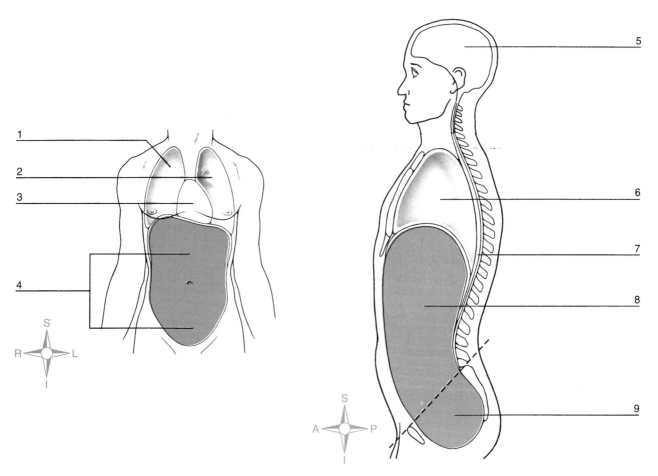

Figure 8.3   Label the names of these body cavities.

Figure 8.4   Label the nine regions of the abdominopelvic cavity.

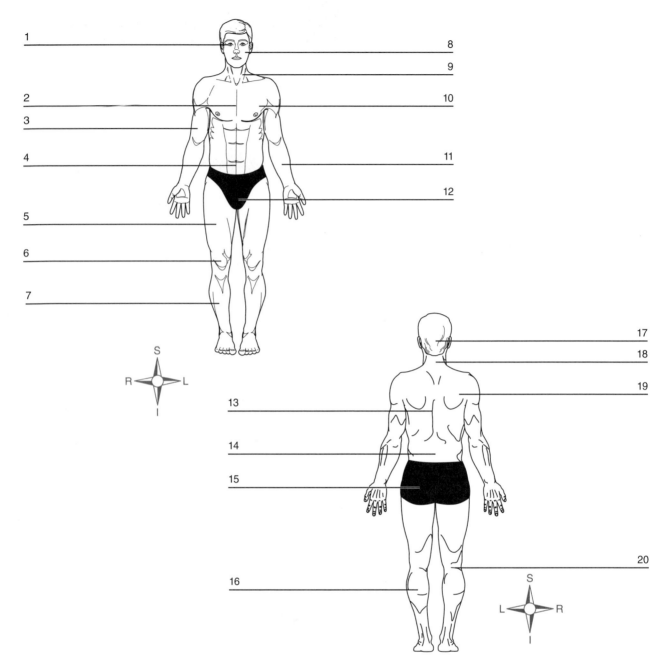

Figure 8.5 Label these figures using the regional terms listed in activity D.

○ **buccal**—cheek (side of mouth)
○ **cervical**—neck
○ **antebrachial**—lower arm
○ **femoral**—upper leg (thigh)
○ **orbital**—eye
○ **patellar**—anterior knee joint
○ **pubic**—lower front of trunk, between legs
○ **thoracic**—chest
○ **tibial**—anterior lower leg (shin)

□ **2** Identify these regions on the posterior aspect of your subject:

○ **calf**—posterior lower leg
○ **cervical**—neck
○ **gluteal**—buttocks
○ **lumbar**—lower back
○ **occipital**—posterior of head
○ **popliteal**—posterior knee joint
○ **scapular**—shoulder blade
○ **thoracic**—upper back

E. Body Systems

As you know, the human organism is composed of organ groups called *systems*. The organs of a system work together in an organized manner to accomplish the function(s) of the system.

As an introduction to human body systems, study table 8.2. Each of the systems will be discussed in more detail later in this course.

Review the organization of the body by completing figure 8.6.

TABLE 8.2   *Human Body Systems*

Organ System	Principal Organs	Primary Function(s)
Integumentary	Skin	Protection, temperature regulation, sensation
Skeletal	Bones, ligaments	Support, protection, movement, mineral/fat storage, blood production
Muscular	Skeletal muscle, tendons	Movement, posture, heat production
Nervous	Brain, spinal cord, nerves, sensory organs	Control/regulation/coordination of other systems, sensation, memory
Endocrine	Pituitary gland, adrenals, pancreas, thyroid, parathyroids, other glands	Control/regulation of other systems
Cardiovascular	Heart, arteries, veins, capillaries	Exchange and transport of materials
Lymphatic	Lymph nodes, lymphatic vessels, spleen, thymus, tonsils	Immunity, fluid balance
Respiratory	Lungs, bronchial tree, trachea, larynx, nasal cavity	Gas exchange, acid-base balance
Digestive	Stomach, intestines, esophagus, liver, mouth, pancreas	Breakdown and absorption of nutrients, elimination of waste
Urinary	Kidneys, ureters, bladder, urethra	Excretion of waste, fluid and electrolyte balance, acid-base balance
Male reproductive	Testes, vas deferens, prostate, seminal vesicles, penis	Continuation of genes (reproduction)
Female reproductive	Ovaries, uterine tubes, uterus, vagina	Reproduction, nurturing of offspring

Anatomical Planes and Directions

PLANES
SAGITTAL 1
MIDSAGITTAL 2
FRONTAL 3
HORIZONTAL 4

DIRECTIONS
LATERAL 5
MEDIAL 6
ANTERIOR 7
POSTERIOR 8
SUPERIOR 9
INFERIOR 10
PROXIMAL 11
DISTAL 12

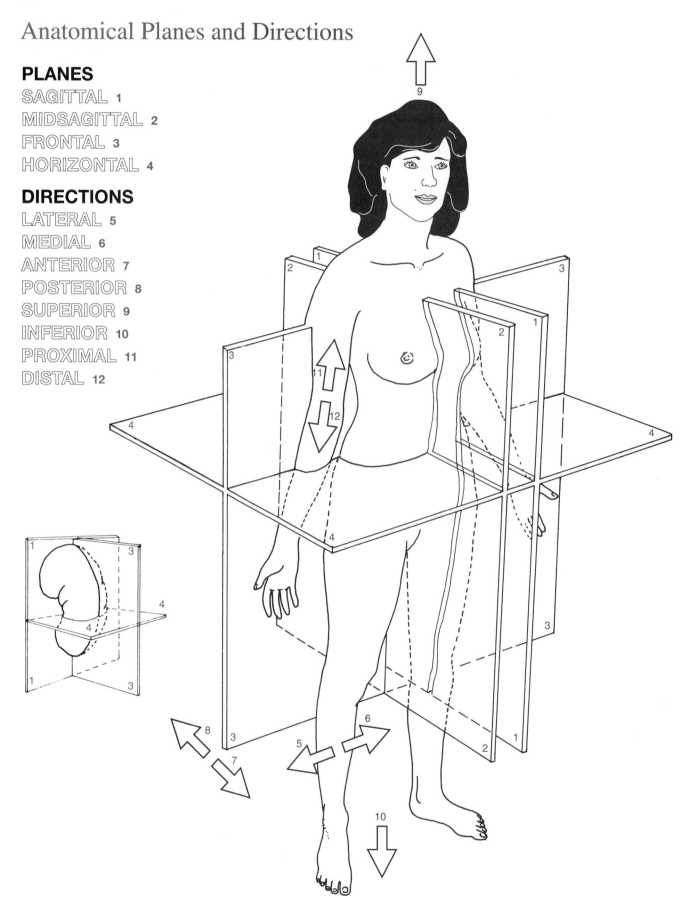

Figure 8.6 Using colored pens or pencils, shade in the figure and accompanying labels as indicated.

LAB REPORT 8

Organization of the Body

Name _____

Date _____   Section _____

Figure 8.3 *(from p. 51)*

_____ 1

_____ 2

_____ 3

_____ 4

_____ 5

_____ 6

_____ 7

_____ 8

_____ 9

Figure 8.4 *(from p. 51)*

_____ 1

_____ 2

_____ 3

_____ 4

_____ 5

_____ 6

_____ 7

_____ 8

_____ 9

Multiple choice

_____ 1

_____ 2

_____ 3

_____ 4

_____ 5

_____ 6

Multiple choice

1. An anatomist cuts a cadaver (preserved body) with a large saw in a way that divides the cadaver into equal left and right halves. The cut is along a __?__ plane.
    - a. sagittal
    - b. midsagittal
    - c. frontal
    - d. horizontal
    - e. *a* and *b*

2. In many study skulls, the top of the skull can be removed so that inner features can be seen. Along which plane should one cut to open the top of a human study skull?
    - a. sagittal
    - b. coronal
    - c. horizontal
    - d. frontal
    - e. *b* and *d*

3. A surgeon makes an incision medially from the left axillary region, turning inferiorly at the midline and proceeding to the pubic region. The path of the cut can be mapped on the patient's chest as

    a. ⌐   b. ┼   c. ⌐   d. ⌐   e. —   f. /

4. Which of these regions contains the spleen?
    - a. epigastric
    - b. hypogastric
    - c. left hypochondriac
    - d. right hypochondriac
    - e. right lumbar

5. Soccer players often wear shin protectors, which shield the __?__ region of each leg.
    - a. femoral
    - b. tibial
    - c. popliteal
    - d. calf
    - e. gluteal

6. Control and regulation of other systems are primary functions of the
    - a. nervous system.
    - b. cardiovascular system.
    - c. endocrine system.
    - d. urinary system.
    - e. *a* and *c*.

Figure 8.5 (*from p. 52*)

_____ 1

_____ 2

_____ 3

_____ 4

_____ 5

_____ 6

_____ 7

_____ 8

_____ 9

_____ 10

_____ 11

_____ 12

_____ 13

_____ 14

_____ 15

_____ 16

_____ 17

_____ 18

_____ 19

_____ 20

Fill-in

_____ 1

_____ 2

_____ 3

_____ 4

_____ 5

_____ 6

_____ 7

_____ 8

_____ 9

_____ 10

_____ 11

_____ 12

Fill-in

Provide the correct anatomical term.

1. The head is __?__ to the feet.
2. The liver is part of the __?__ system.
3. A leg amputation is likely to involve a(n) __?__ cut, or section, through bone.
4. My lower back, or __?__, is sore.
5. The first finger is __?__ to the hand, no matter which position it is in.
6. The popliteal vein is found in the __?__.
7. The heart is __?__ to the right lung.
8. The shoulder is __?__ to the elbow, no matter how one's arm is held.
9. The skin is __?__ relative to the skeleton.
10. Adipose tissue is often just __?__ to the skin.
11. An occipital scar is on the back of the __?__.
12. The thoracic wall is lined with the __?__ layer of the double-layered pleural membrane.

Sketch

Make a rough sketch of a human figure in the position indicated by the compass rosette and figure legend.

Figure 8.A   Anterior view of the human body.

The Skin

The **skin** is the primary organ of the *integumentary system* and is the largest organ of the body. Forming the outer protective covering of the body, the skin is a continuous sheet of **cutaneous membrane.**

Before You Begin

☐ Read the appropriate chapter in your textbook.

☐ Set your learning goals. When you finish this exercise, you should be able to:
  o describe the major structures of the skin and identify their functions
  o identify important skin structures in a diagram, model, and prepared specimen
  o compare and contrast the features of thick skin and thin skin

☐ Prepare your materials:
  o model or chart of a skin cross section
  o microscope
  o prepared microslides:
    thin skin section
    thick skin section
    hair follicles c.s.

☐ **Carefully** read the directions and safety tips for this exercise before starting any procedure.

A. Basic Skin Structure

First in a model, then in prepared microscopic specimens, identify the elements of skin structure identified in the following steps.

SAFETY FIRST!

Don't forget the rules for safe use of the microscope.

☐ **1** The skin has two distinct layers. The superficial layer is a sheet of keratinized stratified squamous epithelium called the **epidermis.** The epidermis is divided into distinct histological regions, or **strata** (meaning "layers"):
  o **Stratum basale** is the deepest stratum of the epidermis. It consists of a single sheet of columnar cells that continue to divide. As the daughter cells are formed, they are pushed upward, becoming part of the next stratum.
  o **Stratum spinosum** is noted for its multilayer of distorted ("spined") cells. The cells become distorted as they are pushed up from the deeper stratum basale and look somewhat "spiny" when stained and prepared for microscopy. Stratum basale and stratum spinosum together are sometimes called **stratum germinativum.**
  o **Stratum granulosum** is superficial to stratum spinosum. It contains flattened cells pushed up from the deeper strata. As the cells are pushed up through this stratum, they form the protein granules that give it the name *granulosum.* By the time the cells leave this stratum, they have died.
  o **Stratum lucidum** (meaning "light layer") is a very thin layer present only in **thick skin.** Thick skin is found only in high-wear areas, such as the palms and soles. The more flexible **thin skin** is found over most other areas of the body. This stratum's name comes from the fact that it is translucent, allowing light to pass through it easily.
  o **Stratum corneum** is the layer of dead, keratinized tissue already identified in Lab Exercise 5. Stratum corneum is extremely thick in thick skin, providing a great deal of protection. Stratum corneum protects deeper tissues from mechanical injury, the inward or outward diffusion of water and other molecules, and invasion by microorganisms.

☐ **2** The layer of skin deep to the epidermis is a sheet of irregular fibrous connective tissue called the **dermis.** The dermis is usually much thicker than the epidermis. Like most connective tissues, the dermis has a scattering of blood vessels and nerves. The blood vessels supply both the dermis and the epidermis. Because blood cools when it travels through the skin, the body varies the amount of blood sent to the skin to regulate loss of heat by the entire body. The dermis contains many sensory nerve endings. Sensations such as *heat, cold, touch,* and *pressure* are mediated by dermal nerve endings. There are two regions of the dermis:

    ○ The **reticular layer** of the dermis is a thick region of irregularly arranged protein fibers. Most of the fibers are collagenous, but a few are made of elastin.

    ○ The **papillary layer** is the bumpy superficial portion of the dermis attached to the epidermis. The bumps, called **papillae** (meaning "nipples"), form regular rows in thick skin but are rather irregularly arranged in thin skin. For this reason, thick skin can be observed to have distinct ridges, such as fingerprints. These ridges give the hands and feet better gripping ability.

☐ **3** Deep to the skin is a layer of **subcutaneous tissue,** sometimes called the **hypodermis** or **superficial fascia.** Although not a part of the skin, it is often studied along with skin. Subcutaneous tissue is loose, fibrous (areolar) connective tissue that connects the skin to underlying muscles and bone. Some of the areolar tissue has been modified to become adipose tissue. Adipose tissue's protective and insulating characteristics complement the protection and temperature regulation roles of the skin.

Hint . . . Histology Minireference Plate 7 shows light micrographs of skin structures. Plate 7*a* represents thin skin, Plate 7*b* thick skin, Plate 7*c* thin skin with hair follicles and other accessory structures, and Plate 7*d* thin skin with sweat glands.

B. Hair, Nails, and Glands

The skin has a variety of accessory structures, including **hair** and **nails.** Both hair and nails are modified forms of stratum corneum, or keratinized tissue. Hair is a cylinder of compact keratinized material, and a nail is a plate of compact keratinized material. Identify the structures described here in a model and in prepared specimens.

☐ **1** Each hair is formed within a separate **hair follicle.** The follicle is a sheathlike indentation of the epidermis. At the bottom of the follicle, a **hair papilla** covered with stratum germinativum produces the hair. The portion of each hair within the follicle is called the **hair root,** whereas the portion that has been pushed out of the follicle is called the **hair shaft.** The hair has a very dense cortex and a less dense medulla.

☐ **2** Attached to each follicle is an exocrine (ducted) **sebaceous gland.** This gland produces the fatty substance, **sebum,** that coats the hair and skin. Sebum prevents moisture loss and conditions the hair and skin so that it doesn't become brittle and easily broken.

☐ **3** The **arrector pili** muscle is a strap of smooth muscle tissue connecting the side of a follicle to the superficial surface of the dermis (fig. 9.1). When contracted, the muscle pulls the follicle so that it is nearly perpendicular to the skin's surface. This increases the air spaces among the hairs, improving its insulation quality. Contraction of the arrector pili also dimples the epidermis, raising a ridge at the edge of the follicle (a "goosepimple").

☐ **4** The toenail or fingernail is also formed by a modified portion of stratum germinativum. In the case of either hair or nail, this modified tissue is often called **matrix.** A portion of the **nail bed** (skin under the nail) is matrix that produces the nail plate. Part of the matrix may be visible through the nail as a pale crescent, or **lunula.** Nail formation begins under a fold of epidermis. The portion of the nail under the fold is the **root,** and the visible portion is the nail **body.** A **cuticle,** or **eponychium,** may extend from the fold onto the nail body (fig. 9.2).

☐ **5** **Sweat glands** are found in many areas of the skin. They are exocrine glands that produce a watery solution, *sweat,* that coats the skin. Sweat primarily improves heat loss by the skin through evaporation. *Eccrine* sweat glands produce thin, watery sweat in many areas of the body. *Apocrine* sweat glands, found in the axillary and pubic regions, secrete a thicker sweat that is rich in complex organic molecules.

Review what you have learned about skin by completing figure 9.3.

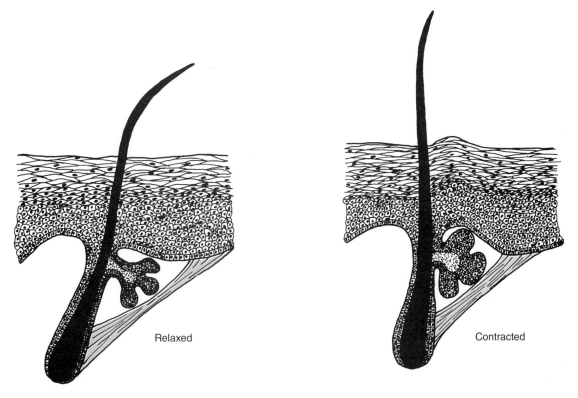

Figure 9.1 When the arrector pili contracts, it pulls the follicle and hair into a perpendicular position, improving the insulation quality of the hair. Notice how a "goosepimple" is raised around the follicle opening.

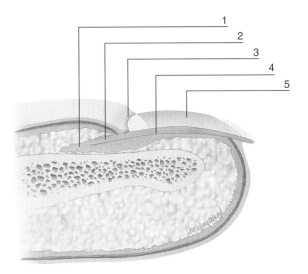

Figure 9.2 Label this section of a finger with the appropriate terms.

The Skin

EPIDERMIS

STRATUM CORNEUM 1
STRATUM LUCIDUM 2
STRATUM GRANULOSUM 3
STRATUM SPINOSUM 4
STRATUM BASALE 5

DERMIS

PAPILLARY LAYER 6
RETICULAR LAYER 7
  SWEAT GLAND 8
  BLOOD VESSEL 9
  NERVE 10

HAIR

ROOT 11
SHAFT 12
FOLLICLE 13
  PAPILLA 14
  MATRIX 15
ARRECTOR PILI 16
SEBACEOUS GLAND 17
HYPODERMIS 18

(a)

(b)

 Figure 9.3 Use colored pens or pencils to color in labels and structures of (*a*) thin skin and (*b*) thick skin.

Clinical Application

The surface area of your skin is a very large boundary between the internal environment of your body and the external environment. The skin is, therefore, a very important line of defense against injury and infection. Unfortunately, the skin barrier is sometimes injured by burns. In clinical situations, it is often useful to estimate the total amount of the skin's surface area that has been damaged. For example, any burn that covers more than 10% of the skin's surface is usually considered to be serious enough to warrant hospitalization. Intravenous (IV) administration of fluids is usually given to any patient with a 15% or greater burn area. Of course, burn surface estimates are important patient statistics that are recorded in each patient's medical record to help health professionals determine *prognoses* (predicted outcomes) and appropriate therapies.

A quick way of estimating burn surface areas is called the **rule of nines** (fig. 9.4). Using this method, one can quickly estimate burns by assigning major regions of the body the surface area value of 9% or a multiple of 9%:

A.	Head and neck,	9%
B.	Upper back torso,	9%
C.	Lower back torso,	9%
D.	Upper front torso,	9%
E.	Lower front torso,	9%
F.	Arm (each),	9%
G.	Front of leg (each),	9%
H.	Back of leg (each),	9%
I.	Perineum (area around genitals),	1%

Using the rule of nines, answer each of the questions that follow.

☐ 1 Baraka was burned while trying to put out a campfire that got out of control. The entire front side of both arms was burned. What is the total surface area of Baraka's burns?

☐ 2 Luke was injured in a cooking accident, which sprayed hot oil all over the front of his face and his chest. What do you estimate to be Luke's total burn surface area? Do you think Luke needs IV fluid therapy?

☐ 3 Johnny was in an automobile accident and was burned badly. The skin of both legs was completely burned, as were Johnny's lower abdomen and the front of his left arm. What is the total surface of Johnny's burn injuries?

☐ 4 Lucy spilled hot water all over her lap, covering the front of her thighs, her genital area, half of her front lower torso, and the front of both arms in burns. What is Lucy's total burn area (in percentage of total skin surface area)? Do you think she will need IV fluid replacement as part of her treatment?

Figure 9.4 Rule of nines for estimating burn surface area of the skin.

LAB REPORT 9

The Skin

Name _____

Date _____ Section _____

Specimen: *thin skin section*   Total magnification: _____

Specimen: *hair follicles c.s.*   Total magnification: _____

Specimen: *thick skin section*   Total magnification: _____

Specimen: _____   Total magnification: _____

Figure 9.2 *(from p. 59)*

_____ 1

_____ 2

_____ 3

_____ 4

_____ 5

Fill-in

_____ 1

_____ 2

_____ 3

_____ 4

_____ 5

_____ 6

_____ 7

_____ 8

_____ 9

_____ 10

_____ 11

_____ 12

_____ 13

_____ 14

_____ 15

_____ 16

_____ 17

_____ 18

_____ 19

_____ 20

Fill-in

1. The ___?___ is the superficial layer (of two) in the skin.
2. The hypodermis is mainly areolar and ___?___ tissue.
3. Fingerprint ridges are formed as a result of the orderly arrangement of dermal ___?___.
4. Stratum corneum is composed of dead cell parts and a tough, waterproof protein called ___?___.
5. Stratum ___?___ is the epidermal stratum in which one could find many phases of mitosis.
6. Cells are pushed into stratum ___?___ from stratum basale.
7. When you sense cold air around you, nerve endings in the ___?___ are probably involved.
8. When you sense cold air around you, bumps may appear on the skin. This is caused by the contraction of the ___?___.
9. The subcutaneous tissue is also called the superficial fascia, or ___?___.
10. Physiologists believe that some sweat glands produce aromatic, organic molecules, called *pheromones,* that may serve as sex attractants. The type of sweat gland (of the two discussed in this exercise) most likely to produce pheromones is ___?___.
11. Strong shampoo tends to remove ___?___ from the hair and scalp, making it dry and easily damaged.
12. The eponychium, or ___?___, covers part of the nail plate.
13. When you have your hair cut, the portion of the hair trimmed is called the ___?___.
14. Blood vessels that supply the epidermis are found in the ___?___ region of the dermis.
15. Stratum basale and stratum spinosum together may be called stratum ___?___.
16. Keratinized ___?___ epithelium forms the epidermis.
17. The dermis is composed mainly of ___?___ fibers, with a few elastic fibers.
18. Stratum ___?___ is usually absent in thin skin.
19. ___?___ glands produce sebum.
20. The most widely distributed type of sweat gland is the ___?___ type of gland.

Table

Fill in the spaces that are blank.

Characteristic	Epidermis	Dermis
Tissue type		
Presence of blood vessels		
Relative thickness		
Permeability		
Relative strength		

Short answer

Write several complete sentences.

How does thin skin differ from thick skin?

LAB EXERCISE 10

Overview of the Skeleton

This exercise is the first of several that concerns the *skeletal system.* The skeletal system's major organs are the **bones, cartilages,** and **ligaments.** Ligaments are simply cords of regular dense fibrous tissue that bind the bones to one another. Bones are more complex in their structure, so we will spend some time investigating the nature of a typical bone. Before we move on to a detailed study of all the bones of the skeleton, we will survey the basic plan of the skeleton.

Investigation of the gross structure of the typical bone and of the basic skeletal plan will be sound preparation for the exercises ahead.

Before You Begin

☐ Read the appropriate chapter in your textbook.

☐ Set your learning goals. When you finish this exercise, you should be able to:
   o describe the organs of the skeletal system
   o list the primary functions of the skeletal system
   o outline the organization of the skeletal system

☐ Prepare your materials:
   o human skeleton (disarticulated)
   o human skeleton (articulated)
   o long bone (fresh, whole)
   o whole long bone (fresh, l.s.)

☐ **Carefully** read the directions and safety tips for this exercise before starting any procedure.

> *Hint . . .* The concepts of this exercise will be easier to understand if you briefly review the microscopic organization of bone tissue presented in Lab Exercise 6 (pp. 35–42).

A. Bone Types

The 206 bones in the standard human skeleton can be classified by their shapes. The best way to learn this classification scheme is by trying to classify the bones yourself, as outlined in the following steps.

☐ 1 Unpack the bones of a disarticulated (taken-apart) human skeleton and spread the bones over your workplace.

☐ 2 Divide the group of bones into four piles, according to the categories given here. Don't use books or other aids to help you.
   o **Long bones** are cylindrical bones that are longer than they are wide.
   o **Short bones** are as long as they are wide, sometimes having an almost cuboidal shape.
   o **Flat bones** arise when bone tissue invades and hardens fibrous membranes, so they are sheetlike in shape. They're usually curved, rather than absolutely flat.
   o **Irregular bones** don't quite fit any of the other categories because of the complexity of their shape.

☐ 3 Compare your results with the results of others in your lab section. Does everyone agree? Compare your results with the samples shown in figure 10.1.

B. Gross Structure of a Bone

All bones have the same general structural pattern. Some bones have more "optional features" than other bones, some less. The long bone is often used as a general specimen for study because it has all the features that any bone can have (fig. 10.2).

SAFETY FIRST!

Because fresh animal tissues at room temperature can harbor dangerous bacterial colonies, specimens should be handled only when wearing disposable, nonporous gloves.

☐ 1 Obtain fresh long bone specimens from a large animal. One should be whole (uncut), the other cut along its long axis (longitudinal section, or l.s.).

☐ 2 Examine the external aspect of a whole bone. Find the features described.
   o **Ligament**—Although actually a separate organ, some bits of these fibrous straps that hold bones together may still be attached to your specimen.

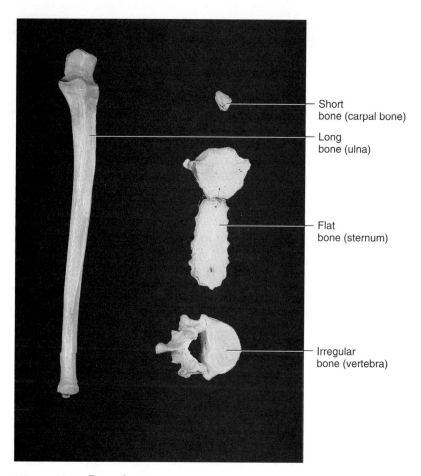

Short
bone (carpal bone)

Long
bone (ulna)

Flat
bone (sternum)

Irregular
bone (vertebra)

Figure 10.1 Bone shapes.

○ **Periosteum**—The periosteum is a sheet of irregular dense fibrous connective tissue continuous with the ligaments. It covers the shaft and part of the heads of a long bone. Try to scrape some of the periosteum away from the underlying bone. How strongly is it attached?

○ **Articular cartilage**—The articular cartilage is a smooth cap of hyaline cartilage found where the bone *articulates* (forms a joint) with another bone. Joints, or connections between bones, are often movable. Which function of the skeletal system benefits by the presence of movable joints?

○ **Diaphysis**—The diaphysis is the whole central shaft of the long bone. Only the external part of the shaft, made of solid bone tissue, is visible from the external aspect. For what skeletal functions is the hard shell of the diaphysis specialized?

○ **Epiphysis**—The epiphyses are the "heads" of a long bone: one proximal to the diaphysis, one distal. Only the external portions are visible in a whole specimen.

□ **3** Use the sectioned bone specimen to identify the structures listed.

○ **Medullary cavity**—The medullary cavity, as its name implies, is a space within the center of the diaphysis. The walls surrounding the space are made of both cancellous and compact bone. The cavity generally contains **yellow bone marrow,** which is a mass of fatty tissue. What is the purpose of yellow marrow?

○ **Endosteum**—Like the periosteum, the endosteum is a sheet of dense fibrous tissue. The endosteum lines the medullary cavity.

○ **Cancellous bone**—The epiphyses have a compact bone cortex, as the diaphysis does, but its medulla is often different. The inside of each epiphysis has cancellous, or spongy, bone. The soft tissue in the spaces of the cancellous bone is red bone marrow, which produces blood cells.

Hint . . . You may want to review the histology of bone and cartilage tissues, as you did in Lab Exercise 6, activity B.

Overview of the Skeleton

LONG BONE
EPIPHYSIS 1
EPIPHYSEAL PLATE 2
DIAPHYSIS 3
PERIOSTEUM 4
ARTICULAR CARTILAGE 5
MEDULLARY CAVITY 6
ENDOSTEUM 7

CANCELLOUS BONE 8
COMPACT BONE 9
BLOOD VESSEL 10

SKELETAL PLAN
AXIAL SKELETON 11
APPENDICULAR SKELETON 12

Figure 10.2 Use colored pens or pencils to color in the labels and the parts of a long bone; then do the same for the skeletal plan.

C. The Plan of the Skeleton

The usual number given for bones in the human skeleton is 206. This is by no means the absolute normal number, however. Most people have more bones, but each person has different types, locations, and numbers of "extra" bones (fig. 10.3). Some people may be missing a bone or two. In this activity, you will examine both the standard 206 bones and the extra bones that may be present.

☐ **1**   Obtain an articulated (put-together) human skeleton.
☐ **2**   The standard **axial skeleton** consists of 80 bones that form the central axis of the skeleton. These 80 bones include 28 skull bones, 1 unattached bone in the throat, 26 vertebrae, and 25 rib cage bones. Locate the bones of the axial skeleton in your specimen. Don't worry about learning the names of individual bones now. That will come later. For now, concentrate on the "big picture" of skeletal organization.

(a)

(b)

Figure 10.3   (*a*) Sesamoid bone at the base of the thumb. (*b*) Sutural (wormian) bones along a suture joint.

SAFETY FIRST!

Be cautious when handling the articulated skeleton. The bones or mounting hardware may become loose and fall from the support frame, injuring you or your lab partners.

☐ **3**   Locate the bones of the **appendicular skeleton** in your specimen. The appendicular skeleton, composed of the 126 nonaxial bones, includes the bones of the appendages, or extremities (arms and legs). Sixty-four of these bones are in the *upper extremities* (shoulders and arms). Sixty-two bones are in the *lower extremities* (hips and legs).
☐ **4**   Ask your lab instructor if there are any extra standard bones or any standard bones missing in your specimen. What difficulties could such differences have caused the individual during life?
☐ **5**   Determine whether your specimen has any of the extra bone types typically found in skeletons:

Hint . . .   A good way to learn the bones and bone markings in the following exercises is to make flash cards. Put the name of the bone or marking on one side of a card and its description or your own sketch or photo on the other side of the card. Then practice *frequently*!

   ○ **Sesamoid bones** are so-called because they resemble sesame seeds: tiny, rounded specks. Sesamoid bones are often found within tendons of the hand and foot (fig. 10.3).
   ○ **Sutural bones,** also called *wormian bones,* are flat bones that form in the sutures (joints) between the cranial bones of the skull (fig. 10.3).

D. Bone Markings and Features

As you have already observed on your specimen, bones do not generally have a smooth surface. There are many bumps, holes, and projections on the bones of the human skeleton. These *bone markings* are named with terms that describe their shape and location. As a preview to the next few exercises, review the terms used to name bone markings, listed in table 10.1. Then use the descriptions and illustrations in your textbook (or later exercises in this lab manual) find an example of each type of bone marking and write them in the last column of Table 10.1 and in the Lab Report.

TABLE 10.1 *Bone Markings*

Marking	Meaning	Example
Angle	Corner	1.
Body	Main portion of a bone	2.
Condyle	Rounded bump; usually fits into a fossa or another bone, forming a joint	3.
Crest	Moderately raised ridge; generally a site for muscle attachment	4.
Epicondyle	Bump near a condyle; often gives the appearance of a "bump on a bump"; for muscle attachment	5.
Facet	Flat surface that forms a joint with another facet or flat bone	6.
Fissure	Long, cracklike hole for blood vessels and nerves	7.
Foramen	Round hole for vessels and nerves (pl., *foramina*)	8.
Fossa	Depression; often receives an articulating bone (pl., *fossae*)	9.
Head	Distinct epiphysis on a long bone, separated from the shaft by a narrowed portion (or neck)	10.
Line	Similar to a crest but not raised as much (is often rather faint)	11.
Margin	Edge of a flat bone or flat portion of an irregular bone	12.
Meatus	Tubelike opening or channel (pl., *meati*)	13.
Neck	Narrowed portion, usually at the base of a head	14.
Notch	V-like depression in the margin or edge of a flat area	15.
Process	Raised area or projection	16.
Ramus	Curved portion of a bone, like a ram's horn (pl., *rami*)	17.
Sinus	Cavity within a bone	18.
Spine	Similar to a crest but raised more; a sharp, pointed process; for muscle attachment	19.
Sulcus	Groove or elongated depression (pl., *sulci*)	20.
Trochanter	Large bump for muscle attachment (larger than tubercle or tuberosity)	21
Tubercle	Smaller version of a tuberosity	22.
Tuberosity	Oblong, raised bump; usually for muscle attachment	23.

Overview of the Skeleton

Name _____

Date _____ Section _____

Multiple choice

_____ 1

_____ 2

_____ 3

_____ 4

_____ 5

_____ 6

Multiple choice

1. The inner lining of the medullary cavity is
   a. made of cancellous bone.
   b. called the endosteum.
   c. called the periosteum.
   d. *a* and *c*.

2. Which of these tissues is present in a typical long bone?
   a. blood tissue
   b. cancellous bone
   c. compact bone
   d. dense fibrous tissue
   e. hyaline cartilage
   f. all of the above

3. In the coloring figure of the long bone, the epiphyseal plate is shown. Although not mentioned in the lab manual, can you deduce the reason for its presence?
   a. It is scar tissue from a previous fracture.
   b. It is an area of growth between the epiphysis and diaphysis during bone development.
   c. It is callous tissue from overuse of the bone.
   d. It is the site of a current fracture.

4. The human skeleton functions to
   a. produce blood tissue.
   b. store fat and minerals.
   c. protect vital organs.
   d. allow movement of the body.
   e. provide a supporting framework.
   f. all of the above.

5. Your physician has just informed you that you have 40 bones in your skull. This means
   a. you have the standard number of skull bones.
   b. you have some sesamoid bones in your skull.
   c. you have some sutural bones in your skull.
   d. you are missing some skull bones.

6. Your physician tells you that all of your knee ligaments have been severed. This means that
   a. your leg bones are not being held together very well.
   b. your leg muscles have become separated from the bone.
   c. your patella (kneecap) is fractured.
   d. your femur (thigh bone) is fractured.

10-8

Fill-in

_____ 1
_____ 2
_____ 3
_____ 4
_____ 5
_____ 6
_____ 7
_____ 8
_____ 9
_____ 10
_____ 11
_____ 12
_____ 13
_____ 14
_____ 15

Axial bones

_____ 1
_____ 2
_____ 3
_____ 4
_____ 5
_____ 6
_____ 7

Appendicular bones

_____ 1
_____ 2
_____ 3
_____ 4
_____ 5
_____ 6
_____ 7
_____ 8

Table 10.1

_____ 1
_____ 2
_____ 3
_____ 4
_____ 5
_____ 6
_____ 7
_____ 8
_____ 9
_____ 10
_____ 11
_____ 12
_____ 13
_____ 14
_____ 15
_____ 16
_____ 17
_____ 18
_____ 19
_____ 20
_____ 21
_____ 22
_____ 23

Fill-in

1. Tiny, round specks of bone found in a tendon are often called __?__ bones.
2. The fibrous covering of a long bone is called the __?__.
3. The shaft portion of a long bone is termed the __?__ of the bone.
4. The head region on the end of a long bone shaft is the __?__.
5. The __?__ on the outside of the long bone is made of hyaline cartilage.
6. Yellow bone marrow is made of __?__ tissue.
7. A bone that is as long as it is wide is classified as a(n) __?__ bone.
8. When classified according to shape, the pelvic bone is considered to be __?__.
9. A human skeleton that is taken apart is called a(n) __?__ skeleton.
10. There are __?__ bones in a *standard* human skeleton.
11. The bones of the upper and lower extremities compose the __?__ skeleton.
12. Red bone marrow is associated with __?__ bone tissue.
13. The __?__ cartilage articulates with another bone or bone process.
14. When classified according to shape, the femur is a(n) __?__ bone.
15. The __?__ is the central space of the long bone.

Skeletal plan

Proceed through this list of bones, beginning at the top, and sort them according to how they fit in the skeletal plan (axial or appendicular). You may need to refer to your textbook or later exercises in this lab manual.

femur

fifth metatarsal

humerus

hyoid

mandible

occipital

patella

pelvic (coxal)

radius

rib

sacrum

scapula

second thoracic vertebra

sternum

tibia

The Skull

This exercise is the first of three exercises that challenge you to learn the bones and important markings of the human skeleton.

The **skull** is the superior portion of the *axial skeleton*. For ease of study, the 28 bones of the skull are divided into three categories. The **cranial bones** form a roughly spherical case for the brain called the *cranium*. The **facial bones** include most of the remaining skull bones, except the 6 **auditory ossicles** of the middle ear. In this exercise, all 28 skull bones are presented for study.

Before You Begin

☐ Read the appropriate chapter in your textbook.

☐ Set your learning goals. When you finish this exercise, you should be able to:
  ○ distinguish between the cranial bones and facial bones of the skull
  ○ identify all the bones of the skull and their important markings in a specimen or figure
  ○ describe the important features of a fetal skull
  ○ name the major joints of the skull and identify them in a specimen or figure

☐ Prepare your materials:
  ○ atlas or chart of the skull (or your textbook)
  ○ articulated human skull (removable top)
  ○ articulated fetal skull
  ○ demonstration pointer

☐ **Carefully** read the directions and safety tips for this exercise before starting any procedure.

Hint . . . Use the diagrams and descriptions in your textbook to find the detailed structures listed in the following activities. Don't forget table 10.1 on page 69 of this manual—it will help you figure out what kind of bone marking you are looking for in your specimen.

A. Bones of the Braincase

Eight of the 28 skull bones form the cranium, or braincase, of the skull. Locate, in a specimen and in a figure, each bone and marking listed, noting the distinguishing characteristics of each (figs. 11.1 and 11.2).

As with all bones and markings studied in this course, try to understand the structural and functional relationships with surrounding structures.

SAFETY FIRST!

Use a disposable plastic pipette (dropper) or another soft *demonstration pointer* to study the features of the skeleton. If you use a pen or pencil, you may mark the skull or damage its delicate parts.

☐ **1** The single **frontal bone** forms the anterior third of the cranial dome. Locate these features on your specimen:
  ○ **supraorbital foramen**
  ○ **frontal sinus**
☐ **2** The left and right **parietal bones** form the middle segment of the cranial dome, joined with each other along the midline by the **sagittal suture.** They form a **coronal suture** with the frontal bone and a **lambdoidal suture** with the occipital bone.
☐ **3** The **occipital bone** forms the posterior portion of the cranial dome, curving inferiorly to the base of the cranium. Locate these important features:
  ○ **foramen magnum**
  ○ **occipital condyles**
☐ **4** The paired **temporal bones** are at the sides of the cranium, extending inward to form part of the cranial floor. Find these temporal structures:
  ○ **zygomatic process**
  ○ **mastoid process**
  ○ **styloid process**
  ○ **external auditory canal**
  ○ **petrous portion**
  ○ **squamosal suture**

The Skull

FRONTAL **1**
PARIETAL **2**
OCCIPITAL **3**
TEMPORAL **4**
SPHENOID **5**
ETHMOID **6**
MAXILLA **7**

LACRIMAL **8**
NASAL **9**
PALATINE **10**
INFERIOR NASAL CONCHA **11**
MANDIBLE **12**
ZYGOMATIC **13**
VOMER **14**

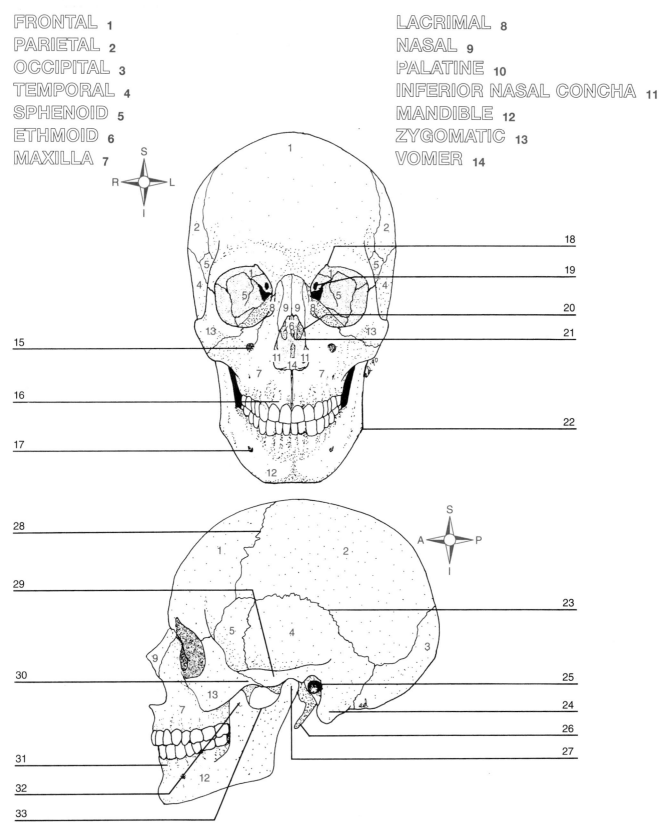

Figure 11.1 Color in the bones of the skull with colored pens or pencils, using the same colors and labels for all skull figures. Write the names of important bone features in the blank label lines.

Figure 11.2 Color in the bones as done in figure 11.1. Fill in the blank labels with the appropriate names.

o **mandibular fossa**
o **temporomandibular joint (TMJ)**
o **internal auditory canal**
o **jugular foramen**
o **carotid foramen**

☐ **5** The **sphenoid bone** is a butterfly-shaped bone that forms part of the anterior floor and sides of the cranium. Locate some of its features:
   o **optic foramen**
   o **superior orbital fissure**
   o **inferior orbital fissure**
   o **foramen rotundum**
   o **foramen ovale**
   o **sella turcica**
   o **sphenoid sinus**

☐ **6** The **ethmoid bone** forms the middle portion of the anterior cranial floor, extending inferiorly between the eye *orbits* to also form the roof of the nasal cavity. Find these important ethmoid structures:
   o **ethmoid sinuses (air cells)**
   o **perpendicular plate**
   o **crista galli**
   o **superior concha**
   o **middle concha**
   o **cribriform plate**
   o **olfactory foramina**

Hint . . . Use the table of bone markings in this lab manual (table 10.1, p. 69) to help you determine the general appearance of each specific feature. Then look for adjectives within the term that help locate the feature. For example, with the term *optic foramen* you know that *foramen* means "hole" and *optic* relates to the eye, so the optic foramen must be a hole related to the eye. In fact, it is the small hole at the back of the eye socket for the optic nerve.

Also use your textbook and anatomical atlases, models, charts, and numbered specimens to help you locate skeletal features.

B. Bones of the Face

Fourteen bones of the skull are classified as facial bones. Find each facial bone and the markings listed in a specimen and figure.

☐ **1** The two **maxillae,** or maxillary bones, are also called the upper jaw bones. They support the face from the eyes down to the mouth, across the front of the cheek. Locate these maxillary features:
   o **palatine portion (process)**
   o **incisive foramen**
   o **alveolar process**
   o **infraorbital foramen**
   o **maxillary sinus**

☐ **2** Each of the two **lacrimal bones** are on the medial margin of an eye orbit, between the ethmoid bone and an upward projection of the maxilla. Each has a sulcus where a tear duct is located (in life).

☐ **3** The left and right **nasal bones** are joined at the midline, forming the superior margin of the nasal opening.

☐ **4** The two **palatine bones** join together at the midline to form the posterior third of the *hard palate* (roof of the mouth). Locate these features:
   o **transverse palatine suture**
   o **greater palatine foramen**

☐ **5** The **inferior nasal conchae,** or turbinate bones, are a pair of thin, curved bones. They project medially from the lateral walls of the nasal cavity, curving toward the nasal floor.

☐ **6** The **mandible,** or lower jaw bone, is an oddly shaped bone forming the lower part of the face. Locate these elements of the mandible:
   o **mandibular arch (notch)**
   o **mandibular condyle**
   o **coronoid process**
   o **alveolar process**
   o **ramus**
   o **angle**
   o **body**
   o **mandibular foramen**
   o **mental foramen**

☐ **7** Each of two **zygomatic bones** form the upper lateral corner of a cheek, from the lower eye orbit around to the temporal bone. Locate
   o **temporal process**
   o **zygomatic arch**

☐ **8** The single **vomer** forms the lower portion of the *nasal septum,* which divides the nasal cavity.

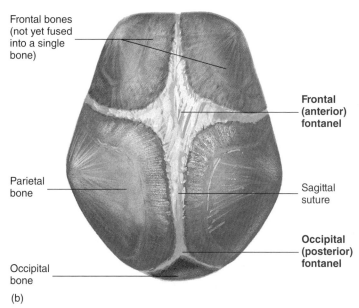

Figure 11.3 Fetal skull showing fontanels and sutures.
(*a*) Lateral view. (*b*) Superior view.

C. Auditory Ossicles

Each middle ear, within the petrous portion of the temporal bone, contains three auditory ossicles. They are listed here, but we will postpone studying them until Lab Exercise 22. (They are not visible in your specimen.)

- ☐ **1** The **malleus** is also called the *hammer.*
- ☐ **2** The **incus** is also called the *anvil.*
- ☐ **3** The **stapes** is known as the *stirrup.*

D. The Fetal Skull

The fetal skull features partly ossified skull bones and a large proportion of fibrous and cartilaginous tissue. Because the flat bones of the cranium have not met to form sutures, there are six fibrous areas called **fontanels.** Identify each in a fetal skull specimen or in fig. 11.3.

- ☐ **1** The two **anterolateral,** or sphenoid, **fontanels** are at the future junction of the sphenoid, temporal, parietal, and frontal bones.

□ **2** The two **posterolateral,** or mastoid, **fontanels** are at the junction of the temporal, parietal, and occipital bones.

□ **3** The **anterior,** or coronal, **fontanel** can be found where the right and left parietal bones are to meet the frontal bone.

□ **4** The **posterior fontanel** forms where the parietal bones are to meet the occipital bone.

LAB REPORT 11

The Skull

Name _____

Date _____ Section _____

Figures 11.1 and 11.2

_____ 15

_____ 16

_____ 17

_____ 18

_____ 19

_____ 20

_____ 21

_____ 22

_____ 23

_____ 24

_____ 25

_____ 26

_____ 27

_____ 28

_____ 29

_____ 30

_____ 31

_____ 32

_____ 33

_____ 34

_____ 35

_____ 36

_____ 37

_____ 38

_____ 39

_____ 40

_____ 41

_____ 42

_____ 43

_____ 44

_____ 45

_____ 46

_____ 47

_____ 48

_____ 49

_____ 50

List the bones of the braincase.

_____ 1

_____ 2

_____ 3

_____ 4

_____ 5

_____ 6

Name the facial bones.

_____ 1

_____ 2

_____ 3

_____ 4

_____ 5

_____ 6

_____ 7

_____ 8

Name the auditory ossicles.

_____ 1

_____ 2

_____ 3

Name the major fontanels.

_____ 1

_____ 2

_____ 3

_____ 4

Sutures

_____ 1

_____ 2

_____ 3

_____ 4

Markings

_____ 1

_____ 2

_____ 3

_____ 4

_____ 5

_____ 6

_____ 7

_____ 8

_____ 9

_____ 10

Anatomical relationships

_____ 1

_____ 2

_____ 3

_____ 4

_____ 5

_____ 6

_____ 7

_____ 8

_____ 9

_____ 10

Sutures

Determine the name of the suture joint described.

1. Joins the parietal bones together
2. Joins the superior margin of the temporal bone to the frontal, parietal, and occipital bones
3. Joins the palatine bone to the maxilla
4. Joins the frontal bone to the two parietal bones

Markings

Give the name of the bone marking or feature described.

1. A large process of the temporal bone, just posterior to the external auditory canal; it contains sinuses
2. A smaller, needle-shaped process just medial to the process described in item 1
3. A hole in the sphenoid bone that allows the optic nerve to exit the eye orbit
4. Same name for ridgelike processes on both the maxilla and mandible in which the teeth are embedded
5. A crest or projection on the superior surface of the ethmoid bone
6. A curved plate of bone projecting from the lateral wall of the nasal cavity, just above the inferior nasal conchae (part of the ethmoid bone)
7. One of two holes on the anterior portion of the mandibular body
8. A hole in the maxilla just below the orbit of the eye
9. Part of the ethmoid that forms the upper portion of the nasal septum
10. Structure formed by both the zygomatic process of the temporal bone and the temporal process of the zygomatic bone

Anatomical relationships

Use a directional term to complete each item correctly.

1. The frontal bone is ___?___ to the occipital bone.
2. The lacrimal bones are on the ___?___ margin of the orbit.
3. The mandible is mostly ___?___ to the maxilla.
4. The occipital condyles are ___?___ to the foramen magnum.
5. The palatine bones are ___?___ to the maxilla.
6. The superior conchae are ___?___ to the nasal septum.
7. The mandibular condyle is ___?___ to the coronoid process.
8. The parietal bones are ___?___ to the frontal bone.
9. The coronal fontanel is ___?___ to the posterior fontanel.
10. The incisive foramen is ___?___ to the transverse palatine suture.

The Vertebral Column and Thoracic Cage

As you recall, the axial skeleton is composed of the skull, the vertebral column, the thoracic cage, and one unattached bone. All these structures form the central core, or *axis,* of the skeleton. In the Lab Exercise 11, we explored the skull in some detail. This exercise presents the remainder of the axial skeleton.

The **vertebral column** is a set of 26 bones stacked one on another to form a slightly curved, flexible support rod. The 25 bones of the rib cage, or **thoracic cage,** protect the lungs and heart within the thorax. The **hyoid bone** is not attached to any other bone but is very close to the skull and vertebral column and lies along the central axis of the body.

Before You Begin

☐ Read the appropriate chapter in your textbook.

☐ Set your learning goals. When you finish this exercise, you should be able to:
  ○ name the component bones of the vertebral column and thoracic cage
  ○ identify the bones and markings of the vertebral column and thorax on a specimen and in figures

☐ Prepare your materials:
  ○ human skeleton (disarticulated)
  ○ human skeleton (articulated)
  ○ human vertebrae set (optional)
  ○ demonstration pointer

☐ **Carefully** read the directions and safety tips for this exercise before starting any procedure.

SAFETY FIRST!

Be careful when handling the skeletal specimens. Loose hardware can injure both you and the specimen. Don't forget to use only approved demonstration pointers.

Hint . . . Use the diagrams and descriptions in your textbook to find the detailed structures listed in the following activities. Don't forget table 10.1 on page 69 of this lab manual—it will help you figure out what kind of bone marking you are looking for in your specimen.

A. The Vertebral Column

The 26 bones of the vertebral column are divided among the bones listed here (figs. 12.1 and 12.2). Find each bone and its important features on both an articulated skeleton and among the separate bones of a disarticulated skeleton or vertebrae set.

☐ **1** The seven **cervical vertebrae** at the superior end of the vertebral column are designated individually by number. The most superior is C1 (cervical number one), the next is C2, and so on. Of C1 through C7, only the first two have commonly used alternate names:
  ○ **Atlas**—C1, or the atlas, is a ringlike vertebra that supports the skull by forming a joint with the occipital condyles. The atlas has an **anterior arch** and **posterior arch** fused to left and right **lateral masses** to form a circle. Flat articulating **facets** are found on both the superior and inferior aspects.
  ○ **Axis**—The axis (C2) is remarkable for its **dens** or *odontoid* (toothlike) process. The dens points superiorly through the atlas to act as a pivot for the rotation of C1 and the skull.
      In an articulated skeleton or figure, notice the **cervical curve** of the vertebral column produced by these seven vertebrae.
☐ **2** The 12 **thoracic vertebrae** (T1 through T12) are inferior to the cervical vertebrae. As a group, they curve in the opposite direction of the cervical curve to form the **thoracic curve.**
☐ **3** Five large **lumbar vertebrae** (L1 through L5) form the curve of the lower back, or **lumbar curve.**

The Vertebral Column

CERVICAL VERTEBRA 1
CERVICAL CURVE 2
THORACIC VERTEBRA 3
THORACIC CURVE 4
LUMBAR VERTEBRA 5
LUMBAR CURVE 6
INTERVERTEBRAL DISKS 7
SACRUM 8
SACRAL CURVE 9
COCCYX 10

Figure 12.1   Color in the labels and structures indicated; then label the bone features on the label lines provided.

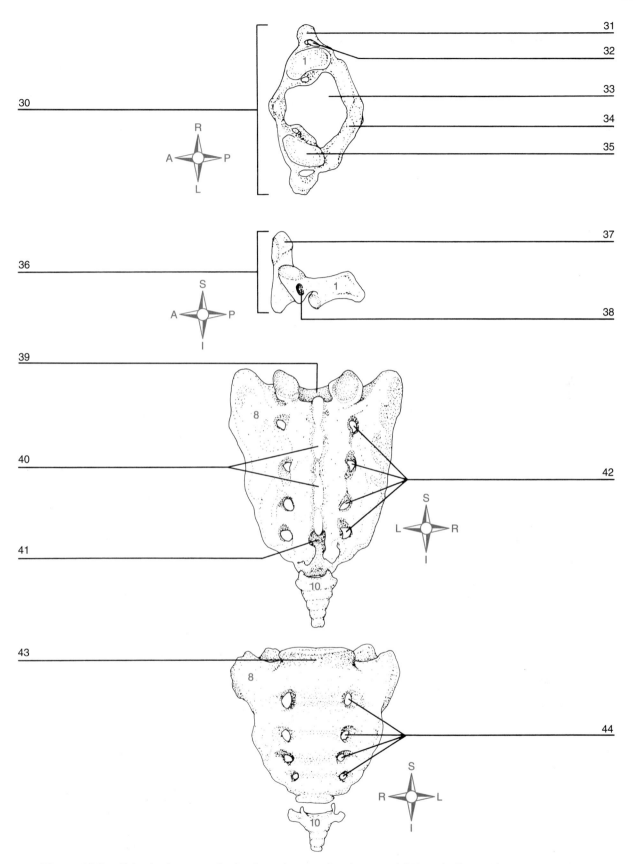

Figure 12.2 Color in these vertebral column bones using the same labels as in figure 12.1; then identify the features indicated.

☐ **4** All vertebrae from C3 to L5 have certain common features. Find each of these features on examples of all three vertebral types.
- **Vertebral foramen**
- **Spinous process (bifid spinous process** in cervical vertebrae)
- **Body**
- **Lamina**
- **Pedicle**
- **Intervertebral foramen**
- **Transverse process** (C1 through C7 also have a **transverse foramen** in each transverse process)
- **Superior articular facets**
- **Inferior articular facets**

☐ **5** The **sacrum** is a bone that develops as a set of five vertebrae that fuse to form one large bone inferior to L5. A slight **sacral curve** can be seen from the lateral perspective. Find these sacral features:
- **Dorsal foramina**
- **Pelvic foramina**
- **Sacral canal**
- **Sacral hiatus**
- **Median sacral crest**
- **Sacral promontory**

☐ **6** The most inferior bone of the vertebral column is the **coccyx,** or tailbone. This bone, like the sacrum, is actually a fusion of several vertebrae.

☐ **7** The **hyoid bone** is not really part of the vertebral column but will be considered here. This bone is a U-shaped bone in the throat. It serves as an attachment for tongue muscles and connective tissue associated with the larynx (voice box). In most articulated skeletal specimens, it is suspended by wire or plastic in a position anterior to the cervical curve.

B. The Thoracic Cage

The 25 bones of the rib cage, or **thoracic cage,** form a partially flexible, protective shield for the heart, lungs, and other thoracic organs (fig. 12.3). The thoracic cage also helps protect some organs of the upper abdomen, such as the liver and spleen. Locate the major structures of the thoracic cage.

☐ **1** Each of the 12 pairs of **ribs** articulates with the vertebral column. The ribs curve around anteriorly. The superior seven pairs of ribs attach directly to the sternum by way of **costal cartilages.** These 14 ribs are also called **true ribs.** The inferior 5 pairs, or **false ribs,** do not directly connect to the sternum. The upper 3 pairs of false ribs indirectly connect to the sternum, via the costal cartilages of true ribs. The last 2 pairs of false ribs don't connect to the sternum at all, so they are also called **floating ribs.** The costal cartilages are made up of hyaline cartilage.

☐ **2** The **sternum,** or *breastbone,* receives the costal cartilages of the true ribs on the anterior aspect of the thoracic cage. In life, it forms an anterior protective wall over the heart and associated structures. Find these structures of the sternum:
- **body (gladiolus)**
- **manubrium**
- **xiphoid process**

The Thoracic Cage

RIBS
TRUE RIB 1
FALSE RIB 2
COSTAL CARTILAGE 3

STERNUM
MANUBRIUM 4
BODY 5
XIPHOID PROCESS 6
THORACIC VERTEBRA 7

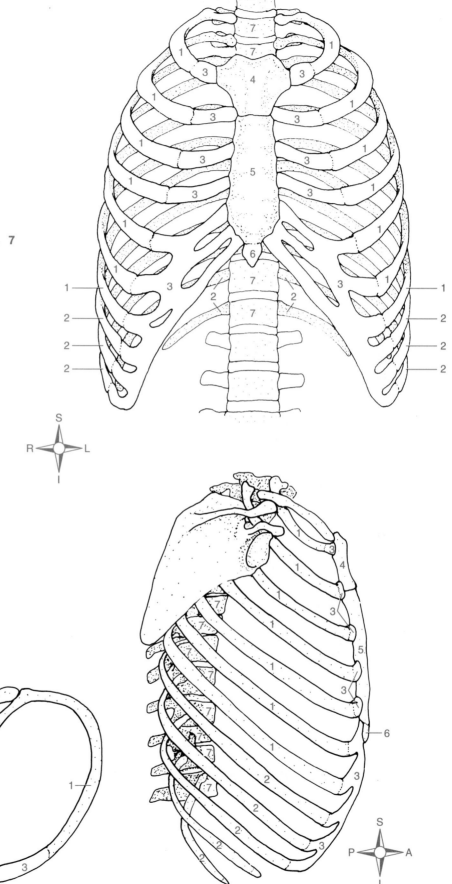

Figure 12.3   Use colored pens or pencils to color in the labels and the bones and bone features of the thoracic cage.

Anatomical Images

Perhaps the oldest method of producing images of internal structures of the human body is sketching dissected specimens. Over the past few decades, biotechnology has produced a number of advanced techniques of producing images of human structures. One such technique, first promoted by Wilhelm Roentgen around the turn of the twentieth century, is **radiography.** This technique, which is also called *x-ray photography,* uses radiation waves that pass through the specimen and onto photographic film. Because the waves pass more easily through soft tissue than through dense tissue, such as bone, shadows of some structures are visible when the film is developed. Figure 12.4*a*

shows a radiograph, or x-ray photograph, of the lumbar region. Can you identify the structures indicated?

A more recent variation of this technique is **computed tomography (CT).** A CT scanner emits a beam of x-rays in a circular path around the subject. After passing through the body, the beams are processed by a computer that is able to reconstruct a three-dimensional image of internal structures. The operator can view different sections of the body on a video monitor. Figure 12.4*b* shows a CT image of a horizontal section of the upper abdomen. A vertebra is clearly shown; can you identify its parts?

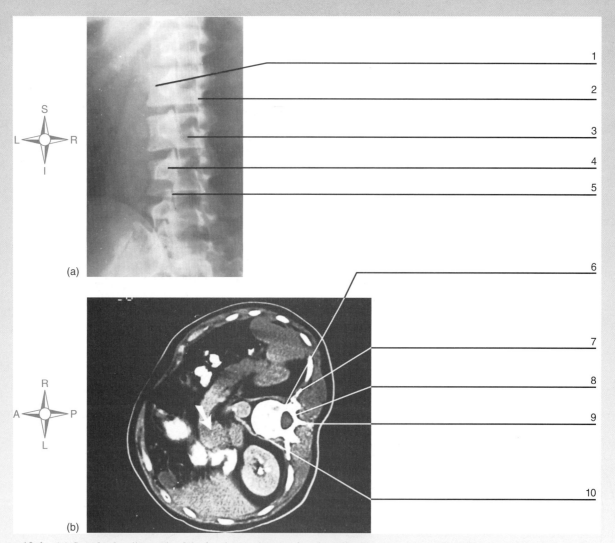

Figure 12.4 (*a*) Standard radiograph of the lumbar region (notice the "tilted" perspective). (*b*) CT image of the abdomen.

The Vertebral Column and Thoracic Cage

Name _____

Date _____ Section _____

Figures 12.1 and 12.2

_____ 11

_____ 12

_____ 13

_____ 14

_____ 15

_____ 16

_____ 17

_____ 18

_____ 19

_____ 20

_____ 21

_____ 22

_____ 23

_____ 24

_____ 25

_____ 26

_____ 27

_____ 28

_____ 29

_____ 30

_____ 31

_____ 32

_____ 33

_____ 34

_____ 35

_____ 36

_____ 37

Table

Fill in the numbers as indicated.

Number	Structures
	Bones in the vertebral column
	Unfused vertebrae
	Cervical vertebrae
	Thoracic vertebrae
	Lumbar vertebrae
	Spinal curves
	Bones in the thoracic cage
	Pairs of ribs
	Pairs of true ribs
	Pairs of false ribs
	Pairs of floating ribs

Figures 12.1 and 12.2 (continued)

_____ 38

_____ 39

_____ 40

_____ 41

_____ 42

_____ 43

_____ 44

Figure 12.4

_____ 1

_____ 2

_____ 3

_____ 4

_____ 5

_____ 6

_____ 7

_____ 8

_____ 9

_____ 10

Fill-in

_____ 1

_____ 2

_____ 3

_____ 4

_____ 5

_____ 6

_____ 7

_____ 8

_____ 9

_____ 10

_____ 11

_____ 12

_____ 13

_____ 14

_____ 15

_____ 16

_____ 17

_____ 18

_____ 19

_____ 20

Fill-in

1. The first cervical vertebra is called C1, or ___?___.
2. The second cervical vertebra is called C2, or ___?___.
3. The inward curve of the lower back is called the ___?___ curve.
4. All ribs articulate with ___?___.
5. Both radiographs and CT scans use ___?___ to form images of anatomical structures.
6. Kyphosis is a condition in which the vertebral column is distorted, giving a person a humpback. This condition involves an exaggerated ___?___ curve.
7. Both the coccyx and ___?___ are vertebral bones that are actually fused vertebrae.
8. The thoracic vertebrae are ___?___ to the lumbar vertebrae.
9. The superior portion of the sternum is called the ___?___.
10. The only bone that does not articulate with another bone is the ___?___.
11. The ___?___ is a vertical tunnel through the sacrum for the passage of nerves descending from the spinal cord.
12. The medial bumps along the midline of the back are caused by the ___?___ of each vertebra.
13. The ___?___ ribs do not articulate, even indirectly, with the sternum.
14. The ___?___ ribs connect directly to the sternum.
15. Costal cartilages are composed of ___?___ cartilage tissue.
16. ___?___ disks (see figure 12.1) form a type of joint between the bodies of vertebrae.
17. The ___?___ is a horseshoe-shaped bone.
18. The tailbone is more properly known as the ___?___.
19. Spinal nerves leave the protection of the vertebral column by way of lateral holes, or gaps, between the vertebrae called ___?___ foramina.
20. The medial ridge along the posterior surface of the sacrum is called the ___?___.

LAB EXERCISE 13

The Appendicular Skeleton

Remember that the human skeleton is divided into the axial skeleton and the appendicular skeleton. You are now ready to learn the appendicular skeleton.

The appendicular skeleton consists of all 126 bones that form the upper and lower extremities. The *shoulder girdle* and arms (upper extremities) have 64 bones altogether, whereas the *pelvic girdle* and legs (lower extremities) have a total of 62 bones. This exercise presents all the bones and many important bone features of the appendicular skeleton.

Before You Begin

☐ Read the appropriate chapter in your textbook.

☐ Set your learning goals. When you finish this exercise, you should be able to:
   o name the bones of the appendicular skeleton
   o identify appendicular bones and markings on a specimen and in figures

☐ Prepare your materials:
   o human skeleton (disarticulated)
   o human skeleton (articulated)
   o demonstration pointer

☐ **Carefully** read the directions and safety tips for this exercise before starting any procedure.

SAFETY FIRST!

Beware of loose parts in the articulated skeleton. Remember to use only approved demonstration pointers.

Hint . . . Use the diagrams and descriptions in your textbook to find the detailed structures listed in the following activities. Don't forget table 10.1 on page 69 of this lab manual—it will help you figure out what kind of bone marking you are looking for in your specimen.

A. The Upper Extremities

Find each bone and marking listed in a specimen and in figures (figs. 13.1 and 13.2).

☐ **1** The **scapula,** or shoulder blade, forms part of the shoulder girdle that supports the arms. It is an irregular bone on the superior, posterior aspect of the rib cage that extends laterally to form a joint with the arm. Find these features of the scapula:
   o **axillary border**
   o **vertebral border**
   o **superior border**
   o **acromion**
   o **coracoid process**
   o **glenoid cavity**
   o **spine**

☐ **2** The **clavicle,** or collarbone, is the other bone of the shoulder girdle. The clavicle is a long bone whose long axis lies along a horizontal axis on the anterior, superior aspect of the thoracic cage.

☐ **3** The **humerus** is the large, long bone of the upper arm. Find these parts of the humerus:
   o **head**
   o **anatomical neck**
   o **surgical neck**
   o **greater tubercle**
   o **lesser tubercle**
   o **intertubercular sulcus**
   o **deltoid tuberosity**
   o **coronoid fossa**
   o **olecranon fossa**

☐ **4** The **radius** is one of two long bones of the lower arm. It is lateral to the other lower arm bone (when in the anatomical position), articulating with the capitulum of the humerus. Find these features:
   o **head**
   o **radial tuberosity**
   o **styloid process**

☐ **5** The **ulna,** the other long bone of the lower arm, articulates with the trochlea of the humerus. Find these markings of the ulna:
   o **olecranon**
   o **trochlear notch**
   o **coronoid process**
   o **head**
   o **styloid process**

☐ **6** The **carpus,** or wrist, is composed of eight small bones:
   o **pisiform**
   o **triquetrum (triangular)**
   o **lunate**

○ scaphoid (navicular)
○ hamate
○ capitate
○ trapezoid
○ trapezium

Hint . . . Try to develop a mnemonic sentence to help you remember the location of each carpal bone. A mnemonic sentence is one in which the first letter of each word is the same as the first letter of each carpal bone—in the same order. See p. 149 for an example.

□ **7** The hand consists of five similar **metacarpal bones,** numbered one through five (starting from the thumb side). Articulating with these are the **phalanges,** or finger bones. The phalanges are named *proximal, middle,* and *distal* and are numbered one through five, according to their position. The first middle phalanx (of the thumb) is not present.

B. The Lower Extremities

Find each of the bones and markings listed (figs. 13.3 and 13.4).

□ **1** The left and right **coxal bones,** or pelvic bones, articulate with the sacrum to form the pelvic girdle. The *coxae* articulate with the sacrum and support the legs. The coxae are formed by the fusion of three bones early in development: the **ilium, ischium,** and **pubis.** These bones are still identifiable as pelvic regions. Find these other coxal features:
○ **acetabulum**
○ **iliac crest**
○ **anterior superior spine**
○ **ischial spine**
○ **obturator foramen**
○ **symphysis pubis**

○ **pubic arch**
○ **sacroiliac joint**

□ **2** The **femur** is the long bone of the upper leg, or thigh bone. Find these features of the femur:
○ **head**
○ **neck**
○ **greater trochanter**
○ **lesser trochanter**
○ **medial condyle**
○ **lateral condyle**

□ **3** The **patella** is a large sesamoid bone forming the anterior bone of the knee joint.

□ **4** The **tibia** is one of two long bones of the lower leg. The tibia is the larger, medial lower leg bone. Find these tibial features:
○ **lateral condyle**
○ **medial condyle**
○ **tibial tuberosity**
○ **medial malleolus**

□ **5** The **fibula** is the narrower, lateral bone of the lower leg, having these parts:
○ **head**
○ **lateral malleolus**

□ **6** The seven **tarsal bones** form the ankle:
○ **talus**
○ **calcaneus**
○ **cuboid**
○ **navicular**
○ **medial cuneiform**
○ **intermediate cuneiform**
○ **lateral cuneiform**

Hint . . . Try to develop a mnemonic sentence to help you remember the correct order of the tarsal bones.

□ **7** The foot is composed of five **metatarsal bones,** similar to the metacarpals of the hand. Also like the hand, the foot has 28 **phalanges.**

Radiograph of the Foot

Identify the labeled structures of this x-ray, or radiograph, of the medial view of the left foot.

_____ 1   _____ 5

_____ 2   _____ 6

_____ 3   _____ 7

_____ 4   _____ 8

The Upper Extremity

SCAPULA 1

CLAVICLE 2

HUMERUS 3

RADIUS 4

ULNA 5

PISIFORM 6

TRIQUETRUM 7

LUNATE 8

SCAPHOID 9

HAMATE 10

CAPITATE 11

TRAPEZOID 12

TRAPEZIUM 13

METACARPAL 14

PROXIMAL PHALANX 15

MIDDLE PHALANX 16

DISTAL PHALANX 17

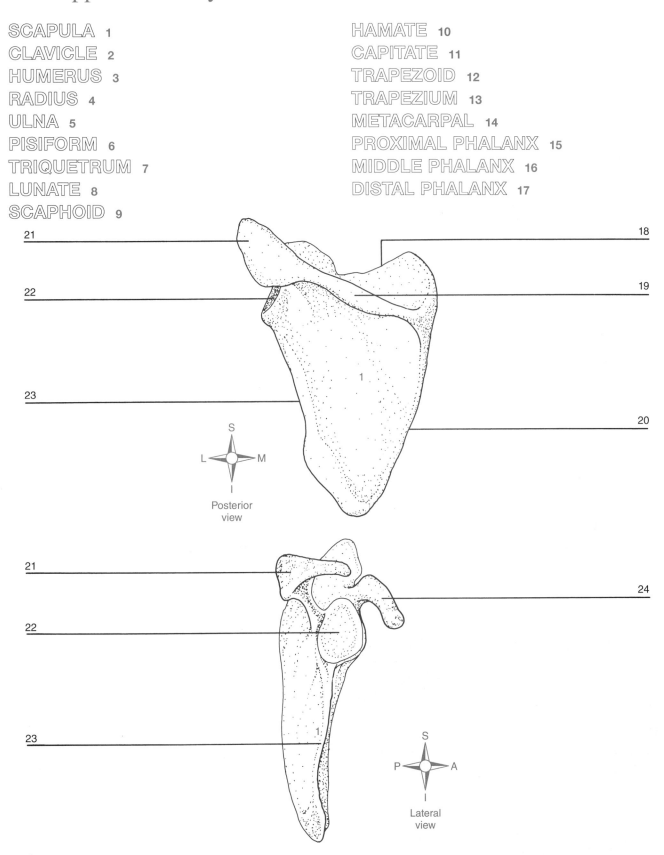

Figure 13.1 Color in the labels and the bones of the upper extremity, and label bone parts as indicated.

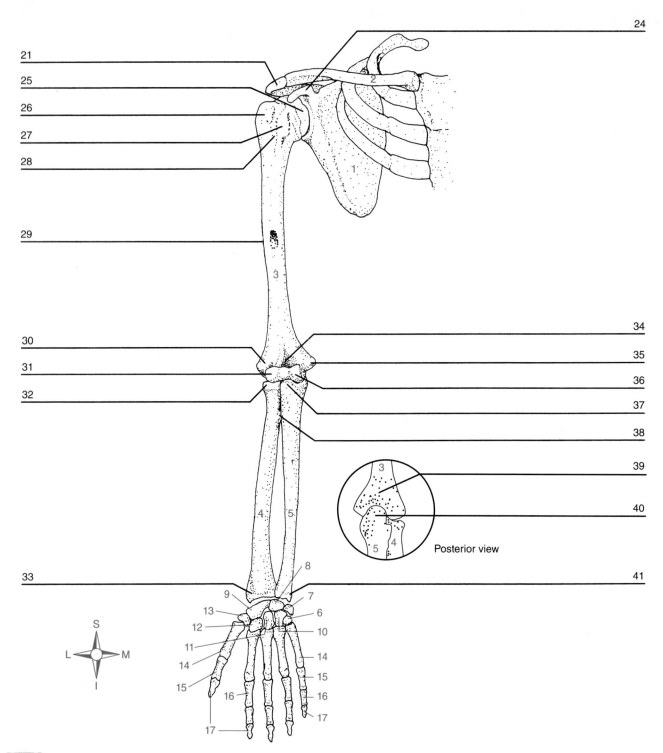

Figure 13.2 Use the labels from figure 13.1 and color in the bones of the upper extremity in this anterior view. The inset is a posterior view of the elbow joint.

The Lower Extremity

COXAL 1
ILIUM 2
ISCHIUM 3
PUBIS 4
FEMUR 5
PATELLA 6
TIBIA 7
FIBULA 8
TALUS 9
CALCANEUS 10

CUBOID 11
NAVICULAR 12
MEDIAL CUNEIFORM 13
INTERMEDIATE CUNEIFORM 14
LATERAL CUNEIFORM 15
METATARSAL 16
PROXIMAL PHALANX 17
MIDDLE PHALANX 18
DISTAL PHALANX 19

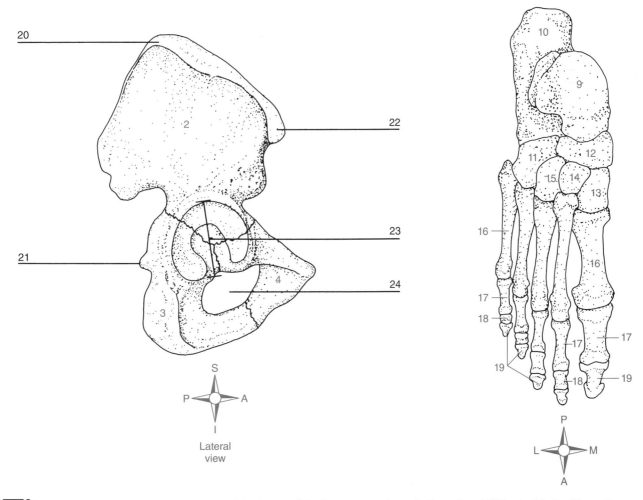

Figure 13.3 Color in the labels and the bones of the lower extremity as indicated, and fill in the labels of bone features.

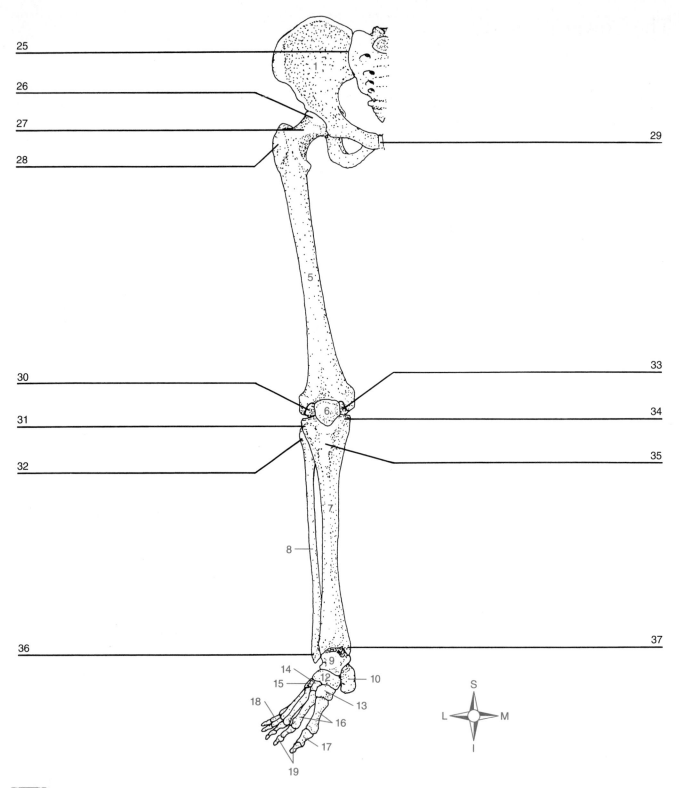

25

26

27

28

29

30

31

32

33

34

35

36

37

Figure 13.4 Use the coloring labels from figure 13.3 and color in the bones. Fill in the labels of bone features.

LAB REPORT 13

Name _____

The Appendicular Skeleton

Date _____ Section _____

Figures 13.1 and 13.2

_____ 18

_____ 19

_____ 20

_____ 21

_____ 22

_____ 23

_____ 24

_____ 25

_____ 26

_____ 27

_____ 28

_____ 29

_____ 30

_____ 31

_____ 32

_____ 33

_____ 34

_____ 35

_____ 36

_____ 37

_____ 38

_____ 39

Multiple choice

_____ 1   _____ 4

_____ 2   _____ 5

_____ 3   _____ 6

Multiple choice

1. The lower leg is composed of the
    a. tibia.
    b. radius.
    c. ulna.
    d. fibula.
    e. *b* and *c*.
    f. *a* and *d*.

2. Lynn has recently lost quite a bit of weight and now notices that a bump is visible in each side of her lower abdomen. Palpating it, she feels that her pelvic bone is protruding. Which part?
    a. ischial spine
    b. symphysis pubis
    c. anterior superior spine
    d. iliac crest
    e. acetabulum

3. The sacrum and coxal bone are joined at the
    a. sacroiliac joint.
    b. hip joint.
    c. sacrocoxal joint.
    d. ischial tuberosity.
    e. symphysis pubis.

4. Jeb accidentally smashed his hand in a car door, breaking one of his bones. If the injury is near the middle of his palm, which type of bone must be involved?
    a. proximal phalanx
    b. metacarpal
    c. metatarsal
    d. distal phalanx
    e. carpal

5. The humeral process that articulates with the ulna is the
    a. trochlea.
    b. capitulum.
    c. lateral epicondyle.
    d. medial epicondyle.
    e. *a* and *b*.

6. The largest long bone of the appendicular skeleton is the
    a. humerus.
    b. tibia.
    c. femur.
    d. ulna.
    e. clavicle.

Figures 13.3 and 13.4

_____ 20

_____ 21

_____ 22

_____ 23

_____ 24

_____ 25

_____ 26

_____ 27

_____ 28

_____ 29

_____ 30

_____ 31

_____ 32

_____ 33

_____ 34

_____ 35

_____ 36

_____ 37

Radiograph (foot)

_____ 1

_____ 2

_____ 3

_____ 4

_____ 5

_____ 6

_____ 7

_____ 8

Table

Fill in the numbers as indicated.

Number	Structures
	Bones in the appendicular skeleton
	Bones in the upper extremities
	Bones in the lower extremities
	Wrist bones (total)
	Ankle bones (total)
	Hand/finger bones (total)
	Foot/toe bones (total)

Joints

Joints, or **articulations,** are structures in the skeleton where two or more bones join together. Because different joints are built differently and therefore behave differently, anatomists have devised a scheme for classifying articulations. This exercise presents a common system of joint classification and challenges you to find examples of joint classes and the different types of movements that they allow.

Before You Begin

☐ Read the appropriate chapter in your textbook.

☐ Set your learning goals. When you finish this exercise, you should be able to:
  o distinguish among the three classes of joints
  o describe examples of each joint class
  o identify different types of motion in joints

☐ Prepare your materials:
  o human skeleton (articulated)
  o synovial joint (model or fresh animal specimen 1.s.)

☐ **Carefully** read the directions and safety tips for this exercise before starting any procedure.

A. Classifying Joints

Find an example of each type of joint described in the following steps. Try to find examples other than those given here.

SAFETY FIRST!

Use caution when handling the articulated skeleton.

☐ **1** **Fibrous joints** are found where fibrous connective tissue tightly binds the articulating bones. Identify these types of fibrous joints:
  o **Suture**—This is a joint between two flat bones, as between the left and right parietal bones (sagittal suture).
  o **Syndesmosis**—Bands of fibrous tissue bind bones, as between the distal ends of the radius and ulna.
  o **Gomphosis**—A fibrous membrane connects each tooth to its socket in a jaw's alveolar process.
☐ **2** **Cartilaginous joints** are formed when a piece of cartilage joins bones.
  o **Synchondrosis**—Hyaline cartilage connects bones (for example, the costal cartilage connection between a rib and the sternum).
  o **Symphysis**—Fibrocartilage forms the joint, as in the symphysis pubis, joining left and right coxae.

Hint . . . Fibrous and cartilaginous joints can be classified according to function, rather than structure. Thus, functional categories of **immovable** and **slightly movable** are sometimes used. Some fibrous joints are immovable, some slightly movable. Likewise, some cartilaginous joints are immovable, others slightly movable.

☐ **3** **Synovial joints** are always **freely movable** (*diarthrotic*) joints. A flexible *joint capsule,* composed of ligaments and other connective structures and lined with a lubricating *synovial membrane,* allows a wide range of movement. Categories of synovial joints are based on the way in which the articulating bones fit together (fig. 14.1):
  o **Gliding joint**—Two flat surfaces slide past each other, as between two carpal bones.
  o **Hinge joint**—As with a door hinge, two bones are joined so that they can move in one plane only, as in the elbow and knuckle joints of the fingers.
  o **Ellipsoid joint**—An oval condyle fits into an oval fossa, allowing movement in two planes, as between the metatarsals and phalanges of the foot.
  o **Pivot joint**—One bone pivots on the axis of another, allowing rotation, as with the atlas and axis.
  o **Saddle joint**—Two saddle-shaped processes fit together to allow movement in two planes, as

Joint Classification

FIBROUS JOINT 1
CARTILAGINOUS JOINT 2

SYNOVIAL JOINT

ARTICULATING BONE 3
ARTICULAR CARTILAGE 4
SYNOVIAL MEMBRANE 5
SYNOVIAL CAVITY 6
CAPSULE 7

SYNOVIAL JOINT TYPES

GLIDING 8
HINGE 9
ELLIPSOID 10
PIVOT 11
SADDLE 12
BALL-AND-SOCKET 13

Figure 14.1 Color in these labels and figures representing joint classification.

between the thumb's metatarsal and the trapezium of the wrist.
- o **Ball-and-socket joint**—A ball-shaped process fits into a rounded fossa, allowing almost unrestricted movement, as between the femur and the acetabulum.

B. Dissection of a Joint

In a model or fresh animal specimen (longitudinal section) of a synovial joint, such as the knee or shoulder, identify the features described. Sketch your observations on a sheet of paper and attach it to Lab Report 14.

SAFETY FIRST!
Because fresh animal tissue at room temperature may harbor bacterial colonies, handle your specimen with gloved hands.

☐ **1** Identify the *articulating bones.* Which joint is this?
☐ **2** Find the *articular cartilage* on the articulating surfaces of the joint. How would you describe their texture? What is the function of articular cartilage?
☐ **3** Locate the *ligaments* and other connective tissues of the joint capsule holding the joint together. Note how strong and how flexible they are.
☐ **4** The synovial membrane lines the joint capsule. What is its consistency? Is any *synovial fluid* still in the *synovial cavity* formed by this membrane? What is this fluid's function?

C. Joint Movement

Although synovial, or diarthrotic, joints can allow a wide range of motion, they are limited by the structure of the joint and surrounding body parts. Demonstrate the types of joint movement listed by performing them yourself.

SAFETY FIRST!
Be careful to avoid accidentally hitting someone as you do this activity. Don't attempt this activity if you have a physical condition that may be worsened by joint movements.

☐ **1** Some skeletal movements involve the movement of body parts relative to a coronal (frontal) plane (fig. 14.2):
- o **flexion**—decreasing the angle of a joint
- o **extension**—increasing the angle of a joint
- o **hyperextension**—moving a joint beyond its normal range, or beyond the anatomical position
- o **protraction**—moving a part anteriorly, along a horizontal plane
- o **retraction**—moving a part posteriorly, along a horizontal plane

Hint

Figure 14.2 Angular movement of the elbow joint. This example shows angular movement of the elbow joint.

☐ **2** Some motions are done relative to a sagittal plane:
- o **abduction**—moving an appendage's distal end away from the midsagittal plane
- o **adduction**—moving an appendage's distal end toward the midsagittal plane
- o **inversion**—moving the foot from the anatomical position (sole downward) to a position in which the sole is facing the midsagittal plane
- o **eversion**—moving the foot from the anatomical position to a position in which the sole faces away from the midsagittal plane
☐ **3** Try these circular movements:
- o **circumduction**—moving the distal end of an appendage in a circle, making a cone-shaped sweep
- o **rotation**—moving a bone on its axis, as if on a pivot or an axle
- o **pronation**—rotating the forearm from the anatomical position (palm forward) to reverse it (palm facing the posterior)
- o **supination**—rotating the forearm from the pronated position back to the anatomical position

Arthography

Radiographic examination of the soft tissues that form joints is called **arthrography.** In one method of arthrography, a **contrast medium** that absorbs x-rays is injected into the joint. The liquid medium coats the cartilage, ligaments, and other soft structures. When a regular radiograph is taken, the coated soft tissues clearly appear. The image is called an **arthrogram.**

In the knee arthrograms shown in figure 14.3, several features that are invisible in a regular radiograph can be seen:
- **bursa**—a "pillow" made of synovial membrane and filled with synovial fluid
- **meniscus**—a piece of cartilage shaped like a disk with a curved surface
- **articular cartilage**—a cartilage coating on the articulating surfaces of bone

Can you identify these three joint structures in the arthrograms in figure 14.3?

Figure 14.3 Arthrograms of the knee joint.

1. What advantages does using contrast medium give when doing radiography?

2. Computed tomography (CT) has been more recently used to study joints. What advantages does the CT method have over the method described here?

LAB REPORT 14

Joints

Name _____

Date _____ Section _____

Figure 14.3

_____ 1

_____ 2

_____ 3

Matching I

_____ 1

_____ 2

_____ 3

_____ 4

_____ 5

_____ 6

_____ 7

_____ 8

_____ 9

_____ 10

Matching II

_____ 1

_____ 2

_____ 3

_____ 4

_____ 5

_____ 6

_____ 7

_____ 8

_____ 9

_____ 10

Matching I

Answer may be used more than once.

a. Fibrous joint
b. Synovial joint
c. Cartilaginous joint

1. Gliding joint
2. Synchondrosis
3. Freely movable
4. Suture
5. Gomphosis
6. Saddle joint
7. Hinge joint
8. Symphysis
9. Mainly hyaline cartilage or fibrocartilage
10. Ellipsoid joint

Matching II

a. Suture
b. Syndesmosis
c. Gomphosis
d. Synchondrosis
e. Symphysis
f. Hinge
g. Gliding
h. Pivot
i. Saddle
j. Ball-and-socket

1. Between bodies of vertebrae
2. Between the distal ends of the tibia and fibula
3. At the base of the thumb's proximal phalanx
4. Between the articular facets of the vertebrae's processes
5. Between the true ribs and sternum
6. Between the talus and lower leg
7. Between C1 and C2, at the dens
8. The shoulder
9. The transverse palatine suture
10. Between the tooth and jaw

Identify

In the following blanks, write the type of motion illustrated in figure 14.4. For example, in #1 is *abduction* or *adduction* shown?

_____ 1

_____ 2

_____ 3

_____ 4

_____ 5

_____ 6

_____ 7

_____ 8

_____ 9

_____ 10

Figure 14.4 Types of motion.

Organization of the Muscular System

The **muscular system** is composed of all the skeletal muscles, here sometimes called *muscle organs* to distinguish them from muscle tissue and muscle cells. Skeletal muscles are attached to the skeleton in ways that usually move skeletal parts when the muscles contract. The chief function of this system, then, is skeletal movement. Secondarily, this system helps in temperature regulation by providing a source of metabolic heat.

This exercise presents some preliminary information about the muscular system. Although we have seen some of the histology of skeletal muscle tissue, we have not yet explored the muscle fiber or the muscle organ. In this exercise, we will do both. We will also take a brief look at how muscle organs are named so that the next exercise, which involves muscle identification, will be easier.

Before You Begin

☐ Read the appropriate chapter in your textbook.

☐ Set your learning goals. When you finish this exercise, you should be able to:
  o describe the structure of the skeletal muscle fiber
  o identify and describe the principal structures of a skeletal muscle organ
  o define the terms *origin* and *insertion*
  o interpret the meaning of muscle names

☐ Prepare your materials:
  o model or chart of a skeletal muscle cell
  o model or chart of a skeletal muscle organ
  o animal muscle specimen (fresh meat section)
  o dissection tools

☐ **Carefully** read the directions and safety tips for this exercise before starting any procedure.

A. The Skeletal Muscle Fiber

Skeletal muscle fibers provide the basic functional capability of the muscle organ because they can contract with great force. Such contraction requires a rather unusual structural pattern in muscle fiber if it is to be efficient. Study these features of the skeletal muscle fiber in a model or figure (figs. 15.1 and 15.2):

☐ **1** The **sarcolemma** is the outer membrane of the long, cylindrical **muscle fiber.** The sarcolemma has a *resting potential,* or electrical charge. This charge temporarily reverses during an *action potential,* or impulse, when the muscle fiber is stimulated.

☐ **2** The sarcolemma dips inward at several points to form internal *transverse tubules* called **T tubules.** An impulse traveling along the sarcolemma can thus also travel inside the fiber.

☐ **3** Inside the fiber are a number of membranous networks similar to ER. They are all part of the **sarcoplasmic reticulum (SR).** The SR receives the impulse from the T tubules, which are nearby, and release calcium ions into the **sarcoplasm** in response.

☐ **4** The calcium ions released from the SR diffuse through the sarcoplasm among parallel bundles of protein **myofilaments.** Each bundle is called a **myofibril** and is composed of an orderly arrangement of **thin filaments** and **thick filaments.**

☐ **5** The myofilaments are arranged in a repeating pattern called a **sarcomere.** When calcium ions react with some of the myofilament molecules, these filaments slide past one another, shortening each sarcomere. Because all the sarcomeres in the cell's myofibrils shorten, the fiber contracts.

☐ **6** A muscle fiber has multiple **nuclei.** The nuclei are against the inside of the sarcolemma.

Capillary
Mitochondrion
Myofibrils
Nucleus

Sarcolemma
Sarcoplasmic reticulum
T tubule

1

2

3

4

5

6

7

Figure 15.1   Use the terms given to identify the parts of a muscle fiber indicated by the label lines.

B. Structure of a Muscle Organ

Each muscle organ has a unique shape and size, but the principal components of each are the same. Identify each of these in a model or chart. Some structures may be visible in a fresh muscle specimen.

SAFETY FIRST!

Use gloved hands and protective eyewear when handling fresh meat at room temperature. You may wish to use dissection tools. If so, be careful to avoid cuts and puncture wounds.

☐ **1**   The entire organ is covered with fibrous connective tissue that forms a sheath called the **epimysium.**

☐ **2**   The fibrous tissue of the epimysium extends inward to form sheaths around bundles, or **fasciculi,** of muscle cells. This inner fibrous sheath is called the **perimysium.**

☐ **3**   The perimysium's fibrous tissue continues inside each fascicle to wrap around each individual muscle fiber. The connective tissue sheath around individual muscle fibers is called the **endomysium.** Just deep to the endomysium is the sarcolemma.

☐ **4**   All the fibrous sheaths are continuous with one another and contain blood vessels that supply the muscle fibers. At the ends of the muscle organ, the muscle fibers stop while the connective sheaths continue. This forms an extension of dense fibrous connective tissue that attaches to the periosteum of a bone (or to another muscle). If the fibrous connection is in the shape of a strap or band, it is called a **tendon.** If it is a broad, flat sheet, it is called an **aponeurosis.**

☐ **5**   Generally, a muscle attaches to the skeleton at both ends. The end that attaches to the more stationary bone is called the **origin.** The other end, or **insertion,** attaches to the bone that moves as the muscle contracts.

C. Naming Skeletal Muscles

Muscle **nomenclature** is the system of naming skeletal muscle organs. The names of muscles, as with many organs, are in Latin rather than in English. By using only a handful of descriptive Latin words, in combinations of two or three,

Skeletal Muscle

TENDON 1
EPIMYSIUM 2
PERIMYSIUM 3
ENDOMYSIUM 4
MYOFIBRIL 5
THIN FILAMENT 6
THICK FILAMENT 7
SARCOMERE 8
A BAND 9
I BAND 10
Z DISK 11

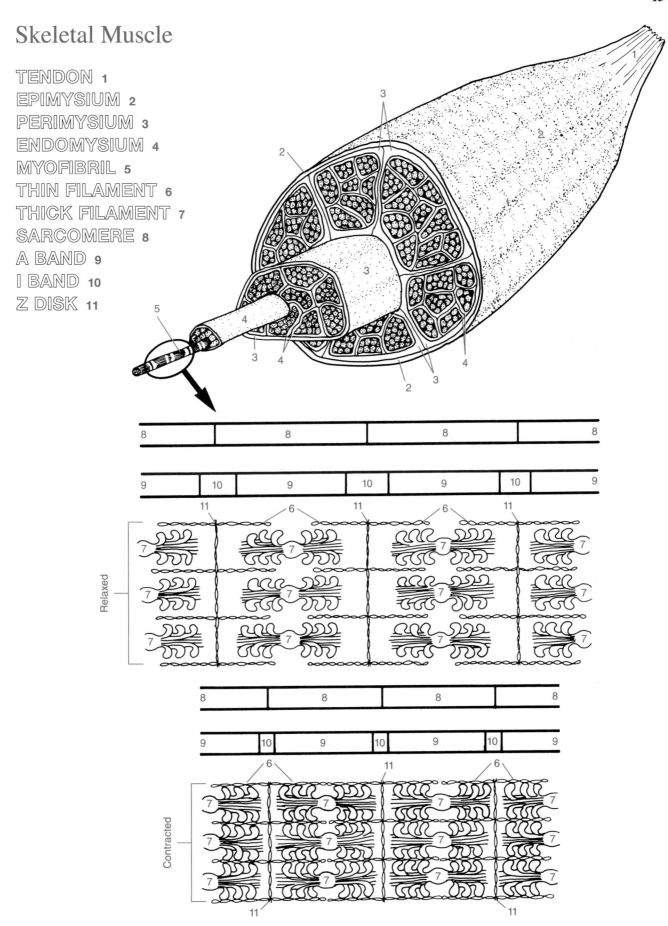

Figure 15.2 Color in these structures and labels associated with skeletal muscle structure.

105

anatomists have named all the muscles of the human body. Review these terms used to name muscles because they will come in handy when you learn specific muscle names in the next exercise.

□ **1** Muscles can be named for their overall shape. Review these terms related to muscle shape:

Deltoid	Shaped like Greek delta (Δ)
Orbicularis	Circular
Platy	Flattened; platelike
Quadratus	Square
Rhomboideus	Diamond-shaped
Trapezius	Trapezoidal
Triangularis	Triangular

□ **2** Some muscles are named for their points of attachment (origin and insertion). For example, the *sternocleidomastoid muscle* has attachments on the sternum, clavicle, and mastoid process of the temporal bone.

□ **3** Muscles can be named according to relative size:

Brevis	Short
Longus	Long
Magnus	Large
Maximus	Largest
Medius	Moderately sized
Minimus	Smallest

□ **4** The direction of fibers visible in a muscle can be a basis for its name, using these terms:

Oblique	Diagonal to the body's midline
Rectus	Parallel to the midline
Sphincter	Circling an opening
Transversus	At a right angle to the midline

□ **5** Some muscle names are derived from the action(s) produced:

Abductor	Abducts a part
Adductor	Adducts a part
Depressor	Depresses a part
Extensor	Extends a part
Flexor	Flexes a part
Levator	Elevates a part
Rotator	Rotates a part

□ **6** Some muscles are named for the region in which they are found. Some of these terms should be familiar to you:

Brachialis	Arm
Femoris	Femur
Frontalis	Frontal (bone)
Gluteus	Posterior of hip/thigh
Oculi	Eye
Radialis	Radius
Ulnaris	Ulna

LAB REPORT 15

Organization of the Muscular System

Name _____

Date _____ Section _____

Figure 15.1 *(from p. 104)*

_____ 1

_____ 2

_____ 3

_____ 4

_____ 5

_____ 6

_____ 7

Fill-in

_____ 1

_____ 2

_____ 3

_____ 4

_____ 5

_____ 6

_____ 7

_____ 8

_____ 9

_____ 10

_____ 11

_____ 12

_____ 13

_____ 14

Put in order

_____ 1

_____ 2

_____ 3

_____ 4

Fill-in

1. The ___?___ is the outer membrane of a muscle fiber.
2. A(n) ___?___ is a bundle of parallel myofilaments within a muscle fiber.
3. The ___?___ is similar to ER but collects and stores calcium ions in a resting muscle fiber.
4. Two types of myofilaments in a skeletal muscle fiber are ___?___ filaments and thick filaments.
5. A repeating pattern, or unit, of myofibrils within a myofilament is a(n) ___?___.
6. A tube formed by the inward extension of the cell membrane is called a(n) ___?___.
7. An entire muscle is covered with a fibrous sheath called a(n) ___?___.
8. A muscle is anchored by its ___?___, whereas it pulls at its insertion.
9. Each fascicle, or fasciculus, is wrapped with a fibrous ___?___.
10. A broad, flat version of a tendon is more accurately called a(n) ___?___.
11. Each muscle fiber has a fibrous ___?___ wrapped around its sarcolemma.
12. A muscle that flexes a joint may have the term ___?___ in its name.
13. A muscle around an opening may be named by the term ___?___.
14. A muscle associated with the femur may have ___?___ in its name.

Put in order

Put these structures in anatomical order from superficial to deep.

sarcolemma

perimysium

endomysium

epimysium

Identify

_____ 1

_____ 2

_____ 3

_____ 4

_____ 5

_____ 6

_____ 7

_____ 8

_____ 9

_____ 10

Identify

Determine the muscle-naming term defined in each item.

1. Associated with the frontal bone
2. Adducts a part
3. Long
4. Encircles an opening or a tube
5. Short
6. Flattened
7. Parallel to the midline of the body
8. Rotates a part
9. Associated with the ulna
10. Extends a part

Muscle Identification

About 36% of the mass in female bodies and 42% of the mass in male bodies is composed of skeletal muscle. In this exercise, some of the larger skeletal muscles of the human body are presented for study.

Before You Begin

☐ Read the appropriate chapter in your textbook.

☐ Set your learning goals. When you finish this exercise, you should be able to:
  o identify the following on a model and in figures:
     muscles of the head and neck
     muscles of the trunk
     muscles of the upper extremity
     muscles of the lower extremity
     muscles of the pelvic floor

  o name the origin and insertion of each major muscle of the body
  o demonstrate the action of each muscle studied

☐ Prepare your materials:
  o models and charts of human musculature
  o demonstration pointers

☐ **Carefully** read the directions and safety tips for this exercise before starting any procedure.

Hint . . . In this exercise, the muscles are presented by their location (head, trunk, etc.). Another common approach is to group muscles by the part they move (muscles that move the head, muscles that move the arm, and so on). Yet another manner of grouping muscles involves their action (flexors, extensors, adductors, etc.). You may want to keep these groupings in mind as you explore the muscular system.

A. Identifying Muscles

Using models and charts of the human musculature, locate the skeletal muscles outlined in the succeeding pages. As you do, note the size and shape of each muscle.

Hint . . . The illustrations shown in this lab exercise are different from those found in your textbook and in any anatomical charts you may have. At first, this will seem unnecessarily frustrating. In fact, however, the differences will help you better learn how to find each muscle in different individual specimens because you'll know where the muscle really is, rather than where a particular label in a particular diagram is located.

☐ **1** Identify these muscles of the *head* and *neck* (table 16.1 and fig. 16.1):
  o **occipitofrontalis** (epicranius)
  o **orbicularis oculi**
  o **orbicularis oris**
  o **buccinator**
  o **zygomaticus** (two muscles)
  o **levator labii superioris**
  o **depressor anguli oris**
  o **temporalis**
  o **masseter**
  o **pterygoids** (two muscles)
  o **sternocleidomastoid**
  o **trapezius**

☐ **2** Identify these muscles of the *trunk* (table 16.2 and fig. 16.2):
  o **erector spinae** (divides into three muscles)
  o **deep back muscles**
  o **scalenes**
  o **external intercostals**
  o **internal intercostals**
  o **rectus abdominis**
  o **external abdominal oblique**
  o **internal abdominal oblique**
  o **transversus abdominis**
  o **trapezius**
  o **levator scapulae**
  o **rhomboids**
  o **serratus anterior**
  o **pectoralis minor**
  o **pectoralis major**
  o **teres major**
  o **latissimus dorsi**
  o **infraspinatus**

TABLE 16.1   *Muscles of the Head and Neck**

Muscle	Origin	Insertion	Action
OCCIPITOFRONTALIS 1	Occipital	Skin of eyebrow, nose	Elevates brows; moves scalp
ORBICULARIS OCULI 2	Maxilla, frontal	Encircles eye, near origin	Closes eye
ORBICULARIS ORIS 3	Maxilla, mandible	Lips	Closes lips
BUCCINATOR 4	Maxilla, mandible	Angle of mouth	Compresses cheeks
ZYGOMATICUS 5 (two)	Zygomatic bone	Angle of mouth, upper lip	Elevate angle of mouth, upper lip
LEVATOR LABII SUPERIORIS 6	Maxilla	Upper lip, nose	Elevates upper lip, nose
DEPRESSOR ANGULI ORIS 7	Mandible	Lower lip near angle	Depresses angle of mouth
TEMPORALIS 8	Temporal aspect of skull	Mandible	Raises jaw
MASSETER 9	Zygomatic arch	Mandible	Raises jaw
PTERYGOIDS (two)	Inferior aspect of skull	Mandible	Medial raises jaw; lateral lowers jaw
STERNOCLEIDOMASTOID 10	Sternum, clavicle	Mastoid process (skull)	Rotates, extends head
TRAPEZIUS 11	Skull, upper vertebral column	Scapula	Extends head, neck

*Use terms in OUTLINE type as coloring labels. Terms in SOLID type do not appear in the coloring plate.

- o **supraspinatus**
- o **subscapularis**
- o **teres minor**
- o **deltoid**

> **Hint . . .** Notice that many of the muscles listed for the trunk include shoulder (upper trunk) muscles that insert on the arm.

☐ **3**   Identify these muscles of the *upper extremity* (table 16.3 and fig. 16.3):
- o **triceps brachii**
- o **biceps brachii**
- o **brachialis**
- o **pronators** (two muscles)
- o **flexor carpi ulnaris**

- o **flexor carpi radialis**
- o **flexor digitorum** (two muscles)
- o **brachioradialis**
- o **supinator**
- o **extensor carpi ulnaris**
- o **extensor carpi radialis** (two muscles)
- o **extensor digitorum**

☐ **4**   Identify these muscles of the *lower extremity* (table 16.4 and fig. 16.4):
- o **iliopsoas**
- o **tensor fasciae latae**
- o **gluteus maximus**
- o **gluteus medius**
- o **quadriceps femoris group: rectus femoris, vastus lateralis, vastus medialis, vastus intermedius**
- o **sartorius**

Muscles of the Head and Neck

Figure 16.1 Color in these skeletal muscles as indicated, referring to the outline labels in table 16.1.

TABLE 16.2   *Muscles of the Trunk**

Muscle	Origin	Insertion	Action
ERECTOR SPINAE 12 (divides into three)	Vertebrae, pelvis	Superior vertebrae, ribs	Holds body upright
DEEP BACK MUSCLES	Vertebrae	Vertebrae	Flex or extend trunk
SCALENES	Cervical vertebrae	First and second ribs	Elevate ribs; expand thorax
EXTERNAL INTERCOSTALS 13	Ribs	Edge of next rib (inferiorly)	Compress thorax
INTERNAL INTERCOSTALS 14	Ribs	Edge of next rib (superiorly)	Expand thorax
RECTUS ABDOMINIS 15	Pubis	Inferior thoracic cage	Flexes waist
EXTERNAL ABDOMINAL OBLIQUE 16	Inferior thoracic cage	Midline of abdomen	Compresses abdomen
INTERNAL ABDOMINAL OBLIQUE 17	Pelvis	Midline of abdomen	Compresses abdomen
TRANSVERSUS ABDOMINUS 18	Vertebrae, pelvis, ribs	Midline of abdomen	Compresses abdomen
TRAPEZIUS 11	Skull, upper vertebral column	Scapula	Extends head, neck; rotates scapula
LEVATOR SCAPULAE 19	Vertebrae	Scapula	Elevates scapula
RHOMBOIDS 20	Vertebrae	Scapula	Retract scapula
SERRATUS ANTERIOR 21	Ribs	Scapula	Protracts scapula
PECTORALIS MINOR 22	Ribs	Scapula	Depresses scapula
PECTORALIS MAJOR 23	Ribs, clavicle	Humerus	Flexes shoulder; adducts arm
TERES MAJOR 24	Scapula	Humerus	Extends shoulder; adducts, rotates arm
LATISSIMUS DORSI 25	Vertebrae	Humerus	Extends shoulder; adducts arm
INFRASPINATUS 26	Scapula	Humerus	Extends, rotates arm
SUPRASPINATUS 27	Scapula	Humerus	Abducts arm
SUBSCAPULARIS	Scapula	Humerus	Extends, rotates arm
TERES MINOR	Scapula	Humerus	Adducts, rotates arm
DELTOID 28	Scapula, clavicle	Humerus	Abducts arm; flexes, extends shoulder

*Use terms in OUTLINE type as coloring labels. Terms in SOLID type do not appear in the coloring plate.

Muscles of the Trunk

Figure 16.2 Color in these muscles associated with the trunk, referring to the outline labels in table 16.2.

TABLE 16.3 *Muscles of the Upper Extremity**

Muscle	Origin	Insertion	Action
TRICEPS BRACHII 29	Humerus, scapula	Ulna	Extends forearm
BICEPS BRACHII 30	Humerus, scapula	Radius	Flexes, supinates forearm
BRACHIALIS 31	Humerus	Ulna	Flexes forearm
PRONATORS 32 (two)	Ulna	Radius	Pronate forearm
PALMARIS LONGUS 61	Humerus (medial epicondyles)	Aponeurosis over palm of hand	Tightens skin of palm
FLEXOR CARPI ULNARIS 33	Humerus (medial epidondyle)	Carpal bone	Flexes, abducts wrist
FLEXOR CARPI RADIALIS 34	Humerus (medial epicondyle)	Metacarpal bones	Flexes, abducts wrist
FLEXOR DIGITORUM 35 (two)	Humerus (medial epicondyle), ulna, radius	Phalanges	Flex fingers
BRACHIORADIALIS 36	Humerus	Radius (distal)	Flexes, pronates forearm
SUPINATOR	Ulna	Radius	Supinates forearm
EXTENSOR CARPI ULNARIS 37	Humerus (lateral epicondyle)	Metacarpal bones	Extends, abducts wrist
EXTENSOR CARPI RADIALIS 38 (two)	Humerus (lateral epicondyle)	Metacarpal bones	Extend, abduct wrist
EXTENSOR DIGITORUM 39	Humerus (lateral epicondyle)	Phalanges	Extends fingers

*Use terms in *OUTLINE* type as coloring labels. Terms in SOLID type do not appear in the coloring plate.*

- o **hamstring group: biceps femoris, semimembranosus, semitendinosus**
- o **adductor group: adductor longus, gracilis**
- o **tibialis anterior**
- o **extensor digitorum longus**
- o **gastrocnemius**
- o **soleus**
- o **peroneus** (three muscles)

☐ **5** Identify these muscles of the *pelvic floor* (table 16.5 and fig. 16.5):
- o **levator ani**
- o **ischiocavernosus**
- o **bulbospongiosus**
- o **gluteus maximus**

B. Demonstrating Muscle Action

For each of the muscles found in activity A of this exercise, demonstrate its action with your own muscle (if possible). As you contract the muscle, palpate it and note its size and location.

SAFETY FIRST!

Be careful to avoid injuring yourself and others as you perform each muscle action.

Muscles of the Upper Extremity

Figure 16.3   Color these muscles of the upper extremity, referring to the outline labels in table 16.3.

115

TABLE 16.4 *Muscles of the Lower Extremity**

Muscle	Origin	Insertion	Action
ILIOPSOAS 40	Ilium, vertebrae	Femur	Flexes hip
TENSOR FASCIAE LATAE 41	Hip	Tibia	Flexes hip; abducts thigh
GLUTEUS MAXIMUS 42	Hip	Femur	Extends hip
GLUTEUS MEDIUS 43	Hip	Femur	Abducts thigh
QUADRICEPS FEMORIS GROUP (a, b, c, d)			
a RECTUS FEMORIS 44	Ilium	Tibia	Extends knee; flexes hip
b VASTUS LATERALIS 45	Femur	Tibia	Extends knee
c VASTUS MEDIALIS 46	Femur	Tibia	Extends knee
d **VASTUS INTERMEDIUS**	Femur	Tibia	Extends knee
SARTORIUS 47	Ilium	Tibia	Flexes hip and knee; rotates thigh
HAMSTRING GROUP (a, b, c)			
a BICEPS FEMORIS 48	Ischium, femur	Fibula	Flexes knee; extends hip
b SEMIMEMBRANOSUS 49	Ischium	Tibia	Flexes knee; extends hip
c SEMITENDINOSUS 50	Ischium	Tibia	Flexes knee; extends hip
ADDUCTOR GROUP (a, b)			
a ADDUCTOR LONGUS 51	Pubis	Femur	Adducts thigh
b GRACILIS 52	Pubis	Tibia	Adducts thigh
TIBIALIS ANTERIOR 53	Tibia	Metatarsal bones	Extends (dorsiflexes) ankle (foot)
EXTENSOR DIGITORUM LONGUS 54	Tibia	Phalanges	Extends toe joints
GASTROCNEMIUS 55	Femur	Calcaneus	(Plantar) flexes ankle (foot)
SOLEUS 56	Tibia, fibula	Calcaneus	(Plantar) flexes ankle (foot)
PERONEUS 57 **(three)**	Tibia, fibula	Tarsal, metatarsal bones	Flex, evert ankle (foot)

*Use terms in *OUTLINE* type as coloring labels. Terms in SOLID type do not appear in the coloring plate.

Muscles of the Lower Extremity

Figure 16.4 Color in these lower extremity muscles, referring to the outline labels in table 16.4.

TABLE 16.5   *Muscles of the Pelvic Floor**

Muscle	Origin	Insertion	Action
LEVATOR ANI 58	Pubis, ischium	Sacrum, coccyx	Elevates anus
ISCHIOCAVERNOSUS 59	Ischium	Clitoris or penis	Compresses base of clitoris or penis
BULBOSPONGIOSUS 60	Male: bulb of penis Female: central tendon of perineum	Male: central tendon of perineum Female: base of clitoris	Male: constricts urethra; assists erection of penis Female: erects clitoris
GLUTEUS MAXIMUS 42	Hip	Femur	Extends thigh

*Use terms in OUTLINE type as coloring labels. Terms in SOLID type do not appear in the coloring plate.

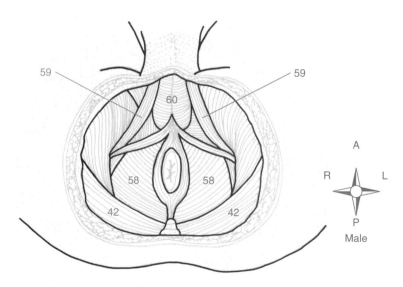

◀◖▬◗▶ **Figure 16.5**   Color in these muscles of the pelvic floor, referring to the outline labels in table 16.5.

Muscle Identification

Name _____

Date _____ Section _____

Table

Use the following muscle names in the table only once.

adductor group

biceps brachii

deltoid

erector spinae

gastrocnemius

gluteus maximus

hamstring group

iliopsoas

latissimus dorsi

pectoralis major

pectoralis major/latissimus dorsi combination

quadriceps femoris group

rectus abdominus

tensor fasciae latae

tibialis anterior

triceps brachii

Part Moved	Flexor	Extensor	Abductor	Adductor
Upper arm				
Lower arm				
Thigh				
Lower leg				
Foot				
Trunk				

Fill-in

_____ 1

_____ 2

_____ 3

_____ 4

_____ 5

_____ 6

_____ 7

_____ 8

_____ 9

_____ 10

_____ 11

_____ 12

_____ 13

_____ 14

_____ 15

_____ 16

_____ 17

_____ 18

_____ 19

_____ 20

_____ 21

_____ 22

_____ 23

_____ 24

_____ 25

Fill-in

1. The occipitofrontalis, or epicranius, originates on the ___?___ bone.
2. The masseter, temporalis, and pterygoids all insert on the ___?___.
3. The sternocleidomastoid inserts on the ___?___.
4. The temporalis raiscs thc ___?___.
5. The ___?___ closes the eye.
6. The pectoralis ___?___ inserts on the humerus.
7. The deltoid inserts on the ___?___.
8. The ___?___ intercostals help accomplish forced expiration.
9. The erector ___?___ help maintain posture.
10. The pectoralis major is ___?___ to the pectoralis minor.
11. The latissimus dorsi is on the ___?___ side of the trunk.
12. The ___?___ intercostals contract when you take a deep breath.
13. The rippling effect seen on the lower midline of the abdomens of some athletes is caused by hypertrophy of the ___?___ muscle.
14. The origin of the brachialis is ___?___ to its insertion.
15. In grasping a baseball tightly in the hand, you would likely use the ___?___ muscle.
16. Both the biceps brachii and the ___?___ flex the forearm.
17. The extensor digitorum inserts on the ___?___.
18. The adductor longus and the ___?___ are both part of the adductor group.
19. The hamstring muscles all ___?___ the leg.
20. The muscle that pulls the leg so that you can cross it over the other leg while sitting is called the ___?___ muscle.
21. Both the gastrocnemius and the ___?___ plantar flex the foot.
22. Muscles of the quadriceps group all ___?___ the leg.
23. In the male, the ___?___ pulls the penis erect but also can constrict the urethra.
24. The large muscle that extends the thigh is the ___?___.
25. The levator ani raises the ___?___.

Muscular Contractions

Now that you are familiar with the structure of the muscular system and its organs, it's time to take a closer look at the physiology of this system. The first activity demonstrates contraction at the cell level. The second and third activities challenge you to investigate concepts of muscle function at the organ level.

Before You Begin

☐ Read the appropriate chapter in your textbook.

☐ Set your learning goals. When you finish this exercise, you should be able to:
  o describe some factors that influence the contraction of single skeletal muscle fibers
  o interpret a myogram of a single twitch, determining its three phases
  o understand the nature of the treppe phenomenon and tetanus
  o demonstrate how muscle contraction can be studied in a laboratory setting

☐ Prepare your materials:
  o glycerinated rabbit psoas muscle
  o watchglass or Petri dish
  o microscopes (dissecting and regular)
  o fine glass probes
  o glass slides and coverslips
  o glycerine solution (50% in triple distilled water)
  o metric ruler (mm)
  o ATP + salt solution
  o ATP solution
  o salt solution
  o PHYSIOLOGY INTERACTIVE LAB SIMULATIONS (PH.I.L.S. 4.0) and computer system (optional)
  If Activity B is performed,
  o frog (live)
  o pithing needle
  o dissection pan or board
  o dissection tools
  o Ringer's solution (amphibian)
  o blunt glass probe or polished tube
  o string or metal S-hook
  o kymograph or physiograph system
  o stimulator apparatus

If Activity C is performed,
  o human myography setup system and manual
  o computer system (CPU, monitor, printer, etc.)
  o electrode gel
  o flat plate electrode (and tape)
  o isolated stimulator
  o pen (washable ink)
  o memory stick(s) (if data are to be stored)

☐ **Carefully** read the directions and safety tips for this exercise before starting any procedure.

A. Contraction of Single Muscle Fibers

In this activity, you will observe contractions in single skeletal muscle fibers. Your specimen is a special glycerinated preparation of a rabbit *psoas muscle.*

SAFETY FIRST!

Take the usual precautions when using a microscope. Protect your eyes and skin from laboratory solutions. Be careful to avoid injury when using the glass probes.

☐ **1** Obtain a small piece (1 to 2 cm) of glycerinated muscle tissue in a watchglass or Petri dish.

☐ **2** Focus on the specimen after placing it on the stage of a dissecting microscope. (You may use a hand lens instead.)

☐ **3** Using fine glass probes or micropipette tips, try to separate some individual fibers from the rest of the bundle. *Avoid using your hands or metal instruments to manipulate the specimen.*

Hint . . . Although it may seem difficult at first to separate fibers, care and patience will get the job done quickly and efficiently. Individual fibers are best, but a few in an unseparated bundle will do.

☐ **4** Remove one of the fibers and place it on the microscope slide with a coverslip. Observe the specimen on a compound microscope.

☐ **5** Observe the muscle fiber with low power, then high power. Make a sketch of your observation. Are the striations visible? Which parts of the muscle fiber form these stripes?

Hint . . . Visualizing the striations, or striping pattern, in glycerinated muscle fibers makes for good practice in honing your skills in microscopy.

Starting at the lowest magnification, turn your light source as low as it will go, then begin reducing the amount of light coming through the diaphragm. See Exercise 1 for details on how to do this.

By trying different light combinations—and having a bit of patience—you will eventually see a faint striated pattern. The stripes are not pigments—they are shadows created by light passing through the regularly arranged filaments that make up the myofibrils within each muscle fiber (see Figure 17.1).

Switch to the high-power objective and adjust the fine focus and light levels to see the striping pattern at higher magnification.

☐ **6** Put three or four separated fibers on a second microscope slide, straight and parallel to one another. Don't use a coverslip. With a metric ruler, measure the length of each fiber (in mm).

☐ **7** With the dropper built into the bottle, use the *ATP + salt solution* to bathe the fibers. Observe the fibers for 30 seconds and note any contraction.

☐ **8** Measure each fiber again and record the new length. Calculate the percentage of change in each fiber using this formula:

$$\frac{(\text{Beginning length}) - (\text{Ending length})}{(\text{Beginning length})} \times 100 = \%$$

What is the *average* percentage of change in the fibers? How do you account for the change? If there was no change, why not?

☐ **9** Repeat steps 6 through 8 using *ATP solution* and again using *salt solution*. Based on your observations, what chemicals are essential for contraction?

Sarcomere

Myofibril

Mitochondria

M line

H zone

Z disk

A band

I band

(a)

A band

I band

A band

H zone

Z disk

Z disk

M line

Z disk

Actin myofilament

Myosin myofilament

(b)   Sarcomere   Sarcomere

Figure 17.1 Striated skeletal muscle appears "striped" because of the regular arrangement of filaments within the microscopic structures of a muscle fiber. (*a*) Muscle fibers (right) at 250× magnification and *electron micrograph* of the arrangement of filaments (left) within them. (*b*) Arrangement of filaments in a pattern of recurring sarcomeres (© Kevin Patton).

Exercise

The term **exercise** has many definitions, depending on the context in which it is used. For some *exercise physiologists,* the term refers to any significant use of skeletal muscles. Exercise physiology has many important applications in athletic training, injury prevention, *ergonomics* (study of body movement as it relates to work activities), physical therapy, and everyday health issues.

An interesting aspect of exercise physiology is that it involves the study of nearly every system and organ in the body, not just the skeletal muscles. When skeletal muscles contract, their sudden increase in metabolism affects many of the processes throughout the body.

For example, the increased use of oxygen in muscles usually triggers an increase in respiratory rate. Why? Because increased oxygen use means more cellular respiration—and that results in more carbon dioxide release into the blood. Respiratory control centers increase the respiratory rate in response to the high blood concentration of carbon dioxide.

Answer the following questions. They will help you appreciate the whole-body aspect of exercise physiology. Read all the questions before answering. You may want to put your answers in the form of a table on a separate sheet.

1. List as many metabolic needs of skeletal muscles during exercise as you can. One has already been given to you: *oxygen.* (Think of things from the extracellular environment that are needed by cells as they work.)

2. For each need listed in question 1, name the body process that supplies that need. Think about *all* the processes that fulfill a need. In our example of oxygen, breathing is cited as a process that supports oxygen availability. However, blood flow is also required to deliver the oxygen.

3. Go over each process listed in question 2. Next to each one, indicate how the process may change during exercise. Does its rate increase? Decrease?

4. During exercise, some processes slow or stop because metabolic necessities, such as oxygen, are diverted to the skeletal muscles. What processes must be "put on hold" until exercise is over?

B. Contraction in a Muscle Organ

Before beginning this activity, review the following basic principles of muscle physiology.

The normal stimulus for a muscle fiber contraction is a *neurotransmitter* chemical from a nerve cell. The nerve-muscle fiber connection is called a **neuromuscular junction.** Electrical or mechanical stimulation can also cause a contraction. Regardless of whether the stimulus is chemical or electrical, it must be strong enough to pass the **threshold of stimulation**— the point at which it is just strong enough to initiate a contraction.

Fibers that constitute a muscle organ are divided into functional teams, or **motor units.** Each fiber in a motor unit is innervated by the same nerve cell and so contracts at the same time. The relative strength of contraction in a whole organ depends in part on how many motor units are stimulated at one time. **Recruitment** of more units increases the organ's strength of contraction.

We will demonstrate muscle organ contraction by stimulating many motor units in a prepared muscle organ. To record our observations, we will use a **myograph** (fig. 17.2a). Myograms, or graphic representations of muscle contractions, can be produced for **isotonic** contractions and for **isometric** contractions. Isotonic contractions change muscle length (without changing tension), and isometric contractions change tension (without changing length). On myograms, time is the horizontal axis, and strength (or length) of contraction is the vertical axis.

The myogram of a single **twitch** contraction shows these features:

o **lag phase**—the phase after stimulation has occurred but during which a contraction is not yet apparent, also called the *latent period*

o **contraction phase**—the period during which the myogram line rises, indicating that contraction is in progress

o **relaxation phase**—the period when the line falls, indicating a return to the resting state

If restimulation of the muscle occurs soon enough after a twitch contraction, you may observe the **treppe phenomenon,** also called the *staircase effect* (fig. 17.2b). The second in a series of contractions has a larger amplitude (size) than the first, the third an even greater amplitude, and so on. Eventually, a maximum is reached, and the waves reach a plateau. When the muscle can no longer sustain contractions, it is in a state of **muscle fatigue.**

When a muscle is continually restimulated before the relaxation phase is over, the myogram will not show the wave line returning to the base line. Instead, the twitch waves will seem to fuse together. Wave summation, or fusion, is often termed **tetanus** (fig. 17.2c). *Incomplete tetanus* occurs when a slight relaxation can be seen between waves. *Complete tetanus* occurs when the waves fuse into a smooth line.

(a)

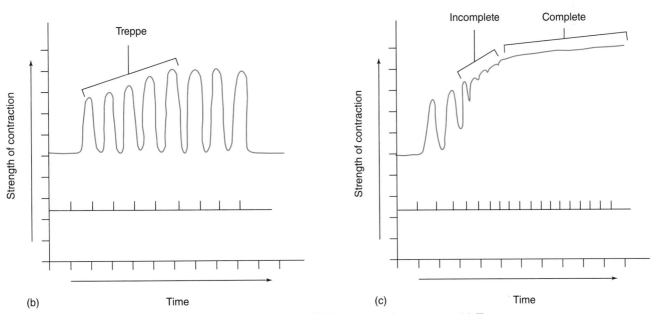

(b)   (c)

Figure 17.2   (*a*) Twitch contractions on a myogram chart. (*b*) The treppe phenomenon. (*c*) Tetanus.

In this activity, you (or your instructor) will use the gastrocnemius muscle of a frog to demonstrate some of the major principles of muscle contraction.

SAFETY FIRST!

Avoid cuts and punctures when using the dissection tools. Check for damaged or improper wiring on the electrical equipment. Do not touch the stimulator electrode or other bare wires directly.

Prepare a frog gastrocnemius muscle in the manner described in steps 1 through 7 (fig. 17.3).

□ **1** Holding the frog in your hand, locate the joint between the base of the skull and the first vertebra by bending the head down until a notch along the posterior edge of the skull can be felt or is visible.

□ **2** Quickly push a pithing needle into the groove, then into the cranial cavity. While rotating the needle, move it from side to side to destroy the brain tissue inside.

□ **3** Partially pull out the needle so that you can insert the tip into the vertebral canal. Push the needle as far as you can, again rotating it and repeatedly moving it up and down to destroy the nerve tissue.

□ **4** Place the frog on a dissection pan or board. Lift the skin around the hip joint with a forceps and puncture it with the tip of a scissors. Try not to damage the muscle tissue of the leg.

□ **5** Using the scissors, cut the skin in a circle all the way around the base of the thigh. Grasp the cut edge of the thigh's skin with forceps and pull it toward the foot. The skin will pull away from the leg as if it were a stocking. Keep pulling until the ankle joint appears.

Hint . . . Keep the muscle tissue moist with Ringer's solution so that it will remain functional.

□ **6** Locate the *gastrocnemius muscle* (the fleshy part of the calf) and the *calcaneal* (Achilles) *tendon,* which attaches it to the ankle. Slip a blunt glass probe under the muscle and slide it back and forth to detach the muscle from underlying connective tissue. While lifting the muscle with the probe, tie a 10 to 15 cm piece of string to the calcaneal tendon (or put a metal S-hook through the tendon as shown in figure 17.4*a*). Finally, cut the tendon distal to the knot.

□ **7** Cut the thigh (bone and muscle) just proximal to the gastrocnemius muscle's attachment to the femur. Break the femur away from the knee joint and remove all muscles but the gastrocnemius. You now have a muscle ready to use in an experiment.

Gastrocnemius

Figure 17.3 Isolating a frog gastrocnemius muscle.

Animals used in research and education should be treated respectfully and humanely. Double pithing is an accepted method for destroying the central nervous system so that the animal will feel no pain. This method also assures that certain tissues will remain functional and won't be damaged by spinal reflexes. The frogs that you use have been raised for research and educational purposes.

□ **8** Set up a kymograph or physiograph as a myograph according to the instructions accompanying the instrument. You may be measuring the intensity of an isotonic contraction by using a movable lever attached to the muscle (as in a kymograph). Or you may be measuring an isometric contraction by means of the force with which it pulls on a transducer (as in a physiograph). Figure 17.4 shows examples of typical setups.

□ **9** Set the stimulator *mode* to CONTINUOUS (REPEAT), *duration* (width) to 1 msec., *frequency* to 1 Hz (1 stimulus per second), and *voltage* to 0.1 volt (v). Stimulate the muscle with the stimulator probe. Continue increasing the voltage in 0.1 v steps until a response is observed. This voltage is the *threshold of stimulation*. Record it here for later reference: _____ volts.

Hint . . . Different systems operate differently, but here's an easy way to calibrate just about any physiograph setup:

- Suspend a 100 gram weight from the transducer.
- Adjust the signal amplifier SENSITIVITY or GAIN to produce a 5 cm deflection of the pen.

Check the manufacturer's handbook for further details regarding your system.

□ **10** Change the stimulator *mode* to SINGLE; then continue to increase voltage over a series of twitches a few seconds apart. You have to trigger each stimulus manually when in the SINGLE mode. Does the amplitude of the myogram curve remain the same? Why or why not? Record the voltage at which a moderately sized twitch is observed: _____ volts.

□ **11** Set the voltage to the level recorded in step 10, the *mode* to CONTINUOUS (REPEAT), and the *frequency* to 1 Hz. Keeping the voltage constant, record a series of twitches 1 second apart. Does the amplitude increase, decrease, or stay the same? Explain. Allow the muscle to rest for several minutes.

□ **12** Keeping the voltage constant, increase the frequency slowly (from 1 Hz) until incomplete tetanus, then complete tetanus, is observed. Record the threshold frequencies for each phenomenon in Lab Report 17.

C. Computerized Myography

This activity may be done in addition to (or in place of) activity B. In this activity, you will use yourself (or your lab partner) as a subject instead of an animal. In this case, a muscle organ is used in vivo (in a living organism) instead of being isolated from the subject's body (fortunately for you). This activity requires the use of any of the many types of available systems. Please use the procedure outlined in the appropriate manual or handout.

SAFETY FIRST!

Don't participate in this activity as a subject if you have health problems, such as a heart condition, that could be affected by this procedure. Be aware of electrical hazards. During the activity, the subject should not touch any person or object other than the pistol grip. Refer to the PHYSIOGRIP manual for additional safety advice.

Advanced Scientific Inquiry

PHYSIOLOGY INTERACTIVE LAB SIMULATION (PH.I.L.S. 4.0)

If available, use the computerized laboratory simulation PH.I.L.S. 4.0 to simulate one or more of the following activities that explore skeletal muscle function:

- *Stimulus-Dependent Force Generation (simulation #4)*
- *Weight & Contraction (simulation #6)*
- *The Length-Tension Relationship (simulation #7)*
- *Principles of Summation and Tetanus (simulation #8)*
- *EMG and Twitch Amplitude (simulation #9)*

For each of the activities that you perform, print the lab report according to the directions given in the activity, and attach a copy of all sheets to your lab report (or as your instructor directs).

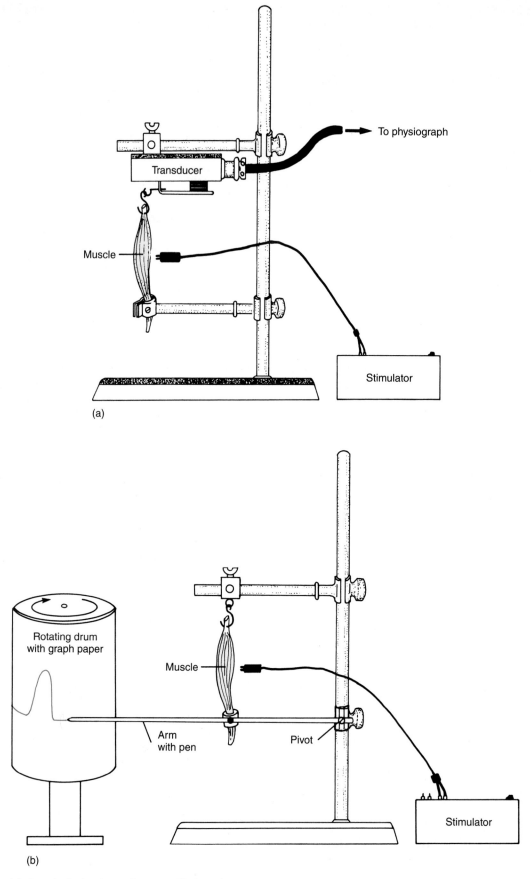

(a)

(b)

Figure 17.4   (*a*) A typical physiograph setup. (*b*) A typical kymograph apparatus. Refer to your system's owner's manual, or your instructor, for detailed instructions in setting up the apparatus.

LAB REPORT 17

Muscular Contractions

Name _____

Date _____ Section _____

Specimen: *psoas muscle (rabbit)*
Total magnification: _____

Specimen: *psoas muscle (rabbit)*
Total magnification: _____

Contraction of Single Muscle Fibers			
Treatment	**Beginning Length**	**Ending Length**	**Percent Change**
ATP + salt solution			
	Average		
ATP solution			
	Average		
Salt solution			
	Average		
Explanation/interpretation:			

Contraction in a Muscle Organ		
Phenomenon	**Value**	**Explanation**
Threshold level of stimulation	_____ volts	What does this value represent?
Level of stimulation required for a twitch	_____ volts	What does this value represent? How does knowing this value help you do the rest of the experiment?
Series of twitch contractions		Does the amplitude remain the same for each twitch in the series? Why or why not?
Threshold frequency for incomplete tetanus	_____ Hz	What does this value represent? A common finding in the frog gastrocnemius is a threshold of about 70 Hz. Is your result higher or lower?
Threshold frequency for complete tetanus	_____ Hz	What does this value represent? A common result is around 200 Hz. How does your result compare?
Duration of lag phase (latent period) (of one twitch)	_____ msec.	What does this value represent? Can you explain it in terms of the physiology of contraction?
Duration of contraction phase (of one twitch)	_____ msec.	What does this value represent?
Duration of relaxation phase (of one twitch)	_____ msec.	What does this value represent?
Total duration of one twitch contraction	_____ msec.	Looking at all your data, does this value seem to be constant or variable?

Note: *Don't forget to attach copies of your myogram charts (with your measurements and notations) to this report.*

Nerves and Reflexes

This exercise is the first of five that challenge you to investigate the human **nervous system.** Before going any further, it is best to lay a foundation by discussing the overall organization of this system.

The nervous system is composed of two major divisions: the **central nervous system (CNS)** and the **peripheral nervous system (PNS).** The CNS includes the brain and spinal cord; the PNS includes all the **nerves** that conduct impulses to and from the CNS. Often, the PNS is further subdivided into the **afferent division** and the **efferent division.** The afferent division includes nerves leading *toward* the CNS, or *sensory nerves.* The efferent division includes nerves leading *away from* the CNS, or *motor nerves.* The efferent division, in turn, can be subdivided into the **somatomotor nervous system** and the **autonomic nervous system.** Although both efferent divisions carry nerve signals away from the brain and spinal cord, they differ in their final destination. Somatomotor nerves end at skeletal muscles, whereas autonomic nerves innervate cardiac muscle, smooth muscle, and glands.

This exercise begins with an investigation of nerve cells and culminates with an exploration of the concept of the reflex arc.

Before You Begin

☐ Read the appropriate chapter in your textbook and Lab Exercise 7.

☐ Set your learning goals. When you finish this exercise, you should be able to:
  o describe the structural components of a typical neuron and identify them on a model or chart
  o outline the features of a reflex arc and be able to apply this model to specific nerve pathways
  o demonstrate nerve reflexes in a human subject

☐ Prepare your materials:
  o model or chart of a multipolar neuron
  o prepared microslide: *spinal cord smear*
  o microscope
  o rubber reflex mallet
  o penlight
  o PHYSIOLOGY INTERACTIVE LAB SIMULATIONS (PH.I.L.S.) 4.0 and computer system (optional)

☐ **Carefully** read the directions and safety tips for this exercise before starting any procedure.

A. The Neuron

As you already know from your study of nervous tissue histology (Lab Exercise 7), the **neuron** is the cell type that conducts impulses, or **action potentials.** The **neuroglia,** on the other hand, support the neurons in any number of ways. Neurons can be *unipolar,* having a single projection from the cell body; *bipolar,* having two projections from the cell body; or *multipolar,* having many projections (fig. 18.1). This activity asks you to find as many of the listed neuron structures as possible on a multipolar neuron model (or chart) and in a slide of multipolar neurons (spinal cord smear).

SAFETY FIRST!
Observe the usual precautions when using the microscope.

☐ 1 Locate the cell body, or **soma.** It is an enlarged area filled with cytoplasm and containing the nucleus and organelles called **Nissl bodies.**

☐ 2 The soma forms a cone-shaped projection, or **axon hillock,** as it projects to become the **axon.** The axon is one of two types of neuron projections (fibers). The axon usually conducts action potentials away from the cell body.

☐ 3 The axon may be wrapped with a series of neuroglial cells called **Schwann cells.** Schwann cells wrap around the axons of some peripheral nerves like tape, each spiraling around a fiber to form a multilayered coating. The inner layers of the Schwann cell are filled with the fatty white substance **myelin.** Because the Schwann cells are found in series, they form a segmented *sheath of Schwann,* or **myelin sheath.** The gaps between the Schwann cells are termed **nodes of Ranvier.** The outer wrapping of each Schwann cell is normal cytoplasm, with organelles and a nucleus. In the case described, the axon is called a **white fiber,** or **myelinated fiber.** A group of white fibers together is called **white matter.** Schwann cells occur only in the PNS. Within the CNS, myelinated axons are wrapped with extensions of **oligodendrocytes,** another type of neuroglial cell.

The Multipolar Neuron

SOMA **1**
  NUCLEUS **2**
    NUCLEOLUS **3**
    MITOCHONDRION **4**
    GOLGI APPARATUS **5**
    NISSL BODY **6**
    AXON HILLOCK **7**
DENDRITE **8**
AXON **9**
SCHWANN CELL **10**

Gemmule

Myelin
sheath of
Schwann cells

Node of
Ranvier

Myelin
sheath

Nucleus of
Schwann cell

Telodendria

Figure 18.1   Use colored pens or pencils to color the labels and this figure of a multipolar neuron.

□ **4** Schwann cells or oligodendrocytes are also associated with **unmyelinated axons,** which together with cell bodies and dendrites form **gray matter.** However, in this case, the neuroglia do not form multiple wrappings and are not partially filled with myelin.

□ **5** **Collateral axons,** or axon branches, can be observed in some cells. Also, the distal ends of axons are often branched. These smaller, distal branches are termed **telodendria.**

□ **6** Multipolar neurons have many projections from the soma called **dendrites.** Dendrites are branched extensions that are sensitive to stimuli from other cells. Other neurons form an association, or **synapse,** at a bump on the dendrite called a **gemmule.** Stimulation of a dendrite results in a local change in potential.

B. Organization of Nerve Pathways

An often used model of nerve pathways is the **reflex arc.** A reflex arc is a way of visualizing the direction of transmission of nerve signals (action potentials). For each step that follows, find the reflex arc featured in figure 18.2.

□ **1** The arc begins with a **receptor,** a specialized cell or cell projection, which is stimulated by a change in its environment. For example, some receptors in the skin are sensitive to heat, others to pressure, and so on.

□ **2** If stimulation of the receptor is significant enough to initiate an action potential in the **afferent neuron,** the signal is transmitted toward the CNS. The afferent, or sensory, neuron brings the signal into the brain or spinal cord.

□ **3** At the peak of the arc, where the signal is "turned around," the afferent neuron may synapse directly

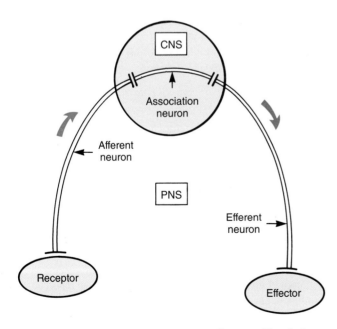

Figure 18.2   A simple three-neuron reflex arc. The dark arrows indicate the direction of nerve transmission.

with an **efferent neuron** (forming a **two-neuron arc**). Often, an **association neuron** synapses with the afferent neuron and transmits the signal to an efferent neuron. This is a **three-neuron arc.**

□ **4** The efferent, or motor, neuron then proceeds to an **effector.** An effector is a muscle or gland innervated by a motor nerve. The effector responds in some way to nerve signals, perhaps by contracting or secreting a chemical.

Synthesizing the component parts of the reflex arc, we see that a stimulus at the receptor results in a reaction by an effector. A simple example is a pin prick that causes a reflexive withdrawal of the pricked limb. Reflexes may be more complex, with sensory information being relayed to several points in the CNS before triggering a motor response. Also, more than one sensory neuron may **converge** on a single association neuron, or multiple motor neurons may **diverge** from a single association neuron. Autonomic reflexes typically involve two efferent neurons, whereas somatomotor reflexes involve only one efferent neuron.

Afferent and efferent neurons generally form bundles of parallel fibers called **nerves** (in the PNS) or **tracts** (in the CNS). Nerves are supported by fibrous connective tissue membranes similar to those found in skeletal muscle organs.

C. Reflex Demonstrations

Examples of reflexes can be seen in clinical tests performed in patients suspected of having some type of nerve damage. If the reflex shows an abnormal reaction or no reaction at all, damage to a component of the reflex arc is suspected.

□ **1** The **patellar reflex,** or *knee-jerk reflex,* is mediated by a two-neuron arc centered in the spinal cord (fig. 18.3*a*). Have the subject sit on a table with the legs dangling above the floor. Sharply tap the knee with a reflex mallet at the ligament just inferior to the patella (kneecap). The tap stretches a quadriceps muscle in the anterior thigh, stimulating *stretch receptors* located in the muscle. In response to the increased stretch, which normally would occur only when the muscle load had suddenly increased, the muscle contracts and extends the lower leg. In this demonstration, such a reaction seems strange, but in normal circumstances the patellar reflex allows the quadriceps muscle to reflexively increase its strength of contraction in response to an increased load.

□ **2** The **triceps reflex** is another stretch reflex (fig. 18.3*b*). Have the subject lie on a table, with an arm across the abdomen. Supporting the subject's arm with the elbow flexed at a 90° angle, sharply tap the posterior surface of the upper arm just proximal to the olecranon. The lower arm should extend as the triceps brachii muscle reflexively contracts. How do you explain this response?

□ **3** The **plantar reflex** involves cutaneous (skin) receptors rather than deep receptors in muscles or

To brain

Afferent
neuron

Muscle
spindle

Motor neuron

Motor endplate

(a)

(b)

(c)

Figure 18.3 (*a*) Patellar reflex arc. The arrow indicates the tapping point. (*b*) Tapping point for the triceps reflex. (*c*) Demonstrating the plantar reflex.

Hint . . . The following are possible abnormal spinal reflex results:

- **Hyperflexia** (exaggerated response) results from damaged or diseased motor areas in the CNS.
- **Hypoflexia** (inhibited response) results from degeneration of nerve pathways, voluntary motor control, and other factors.

tendons (fig. 18.3*c*). Position the subject's bare foot with the lateral surface resting on a table or chair. Demonstrate this reflex by firmly sweeping the handle of the mallet along the lateral region of the sole. In a normal adult, the toes flex. In a **Babinski response,** the toes extend and move apart. The Babinski response is normal for infants in whom the nerves have not fully myelinated but is abnormal in adults.

☐ **4** In a dimly lit area, demonstrate the **pupillary reflexes,** which are centered in the brainstem. Have the subject stare straight ahead, with a hand held vertically between the eyes. Shine a penlight in the subject's left eye (from 5 to 7 cm away). Normally, the light receptors in the eye receive the bright light and trigger a reflexive response by the muscles in the iris. Does the pupil increase or decrease in diameter? What is the advantage of this response? The right pupil may also exhibit a reflexive response, even though light is not shown on it. Does the right pupil increase or decrease in diameter? What advantage is gained by this *consensual* reflex?

Advanced Scientific Inquiry

Physiology Interactive Lab Simulation (Ph.I.L.S. 4.0)

If available, use the computerized laboratory simulation Ph.I.L.S. 4.0 to simulate one or more of the following activities that explore nerve function:

- *Resting Potential and External Potassium Concentration (simulation #10)*
- *Resting Potential and External Sodium Concentration (simulation #11)*
- *The Compound Action Potential (simulation #12)*
- *Conduction Velocity and Temperature (simulation #13)*
- *Refractory Periods (simulation #14)*
- *Facilitation and Depression (simulation #16)*
- *Temporal Summation of EPSPs (simulation #17)*
- *Spatial Summation of EPSPs (simulation #18)*

For each of the activities you perform, print the lab report according to the directions given in the activity and attach a copy of all sheets to your lab report (or as your instructor directs).

LAB REPORT 18

Nerves and Reflexes

Name _____

Date _____ Section _____

Fill in this table, summarizing your examination of a multipolar neuron model. Check off each structure as it is identified. For functions, refer to a reference book or your textbook.

Identification	Structure	Function(s)
☐	Soma	
☐	Nissl bodies	
☐	Axon hillock	
☐	Axon	
☐	Schwann cells	
☐	Nodes of Ranvier	
☐	Collateral axon	
☐	Telodendria	
☐	Dendrite	
☐	Gemmule	

Record your data and interpretations of the reflex tests. Be sure to indicate the function each may have.

Reflex	Result Left	Result Right	Explanation/Discussion
Patellar	☐ Normal ☐ Hyperflexia ☐ Hypoflexia ☐ No response	☐ Normal ☐ Hyperflexia ☐ Hypoflexia ☐ No response	
Triceps	☐ Normal ☐ Hyperflexia ☐ Hypoflexia ☐ No response	☐ Normal ☐ Hyperflexia ☐ Hypoflexia ☐ No response	
Plantar	☐ Normal ☐ Babinski response ☐ No response	☐ Normal ☐ Babinski response ☐ No response	
Pupillary	☐ Dilation ☐ Constriction ☐ No response	☐ Dilation ☐ Constriction ☐ No response	

Fill-in

_____ 1

_____ 2

_____ 3

_____ 4

_____ 5

_____ 6

_____ 7

_____ 8

_____ 9

_____ 10

_____ 11

_____ 12

_____ 13

_____ 14

_____ 15

_____ 16

_____ 17

_____ 18

_____ 19

_____ 20

Put in order

_____ 1

_____ 2

_____ 3

_____ 4

_____ 5

Fill-in

1. The soma forms a cone-shaped __?__ as it projects to form an axon.
2. __?__ are small, distal branches of an axon.
3. Either Schwann cells or extensions of __?__ can form myelin sheaths.
4. The gaps between segments of a myelin sheath are called __?__.
5. A group of myelinated fibers may form a region of nerve tissue called __?__ matter.
6. The __?__ neuron connects an afferent neuron to an efferent neuron.
7. A junction between two neurons, or between a neuron and an effector, is called a(n) __?__.
8. A bundle of parallel neurons encased in fibrous connective tissue is called a(n) __?__ (in the PNS).
9. A bundle of parallel neurons in the CNS is called a(n) __?__.
10. A muscle or gland that is regulated by nerve input is termed a(n) __?__.
11. An action potential traveling down a myelinated axon travels __?__ (faster/slower) than in an unmyelinated axon.
12. Unmyelinated nerve tissue is called __?__ matter.
13. In infants, the __?__ response is normal when testing the plantar reflex.
14. When testing the plantar reflex in an adult, extension of the toes may indicate damage to the __?__ somewhere along the reflex arc.
15. In the __?__ reflex, a muscle contracts when the load increases.
16. The center of the patellar reflex is in the __?__.
17. A stretch reflex involving the biceps brachii muscle causes __?__ of the elbow if the biceps muscle is stretched.
18. In the triceps reflex demonstration, the triceps muscle was the effector, and one or more __?__ were the receptors.
19. If the Achilles (calcaneal) tendon is tapped, one would expect the ankle to __?__.
20. When suddenly illuminated with a penlight, the pupil of the eye normally __?__ (dilates/constricts).

Put in order

Put these components of the reflex arc in the order in which nerve signals pass through them.

association neuron

effector

motor neuron

receptor

sensory neuron

The Spinal Cord and Spinal Nerves

As you know, the spinal cord is one of two principal organs of the central nervous system (CNS). The brain is the other. This exercise presents the spinal cord and its accessory structures on the gross level and the microscopic level. This exercise also includes a peek into the peripheral nervous system (PNS) by means of an introduction to **spinal nerves.**

Before You Begin

☐ Read the appropriate chapter in your textbook.

☐ Set your learning goals. When you finish this exercise, you should be able to:
  o name and describe the coverings of the spinal cord
  o identify principal structures of the spinal cord in figures and specimens
  o define *spinal nerve, nerve root,* and *plexus*
  o name the spinal nerves and locate them in a model or figure

☐ Prepare your materials:
  o preserved specimen: *ox spinal cord segment*
  o dissection tools and tray
  o model or chart of the human spinal cord
  o microscope
  o prepared microslide: *mammalian spinal cord c.s.*

☐ **Carefully** read the directions and safety tips for this exercise before starting any procedure.

A. Spinal Cord Dissection

Use both the preserved spinal cord specimen and a model or chart to explore the features described in this activity.

SAFETY FIRST!

Heed the precautions that accompany the chemical preservative used with your specimen, especially those regarding ventilation and skin/eye protection. Use your dissection tools carefully.

The spinal cord is nerve tissue continuous with that of the brain, extending inferiorly from the brain at the level of the foramen magnum (fig. 19.1). The spinal cord is covered by several layers of protective or nourishing tissue:

☐ **1** The *bony covering* of the spinal cord is formed by the vertebral column. The spinal cord lies within the spinal cavity formed by the vertebral foramina of the stacked vertebrae. What is the bony covering of the brain?

☐ **2** The outermost of three *membranous coverings,* or **meninges** (sing., **meninx**), of the spinal cord is the tough, skinlike **dura mater.** The spinal meninges are continuous with the meninges surrounding the brain.

☐ **3** Deep to the dura mater is a much thinner meninx, the spidery **arachnoid.** The space deep to the arachnoid, the **subarachnoid space,** is filled with circulating **cerebrospinal fluid (CSF)** in life. The CSF has been drained from your specimen.

☐ **4** The deepest meninx, the delicate **pia mater,** is a thin, vascular membrane adhering to the surface of the spinal cord.

☐ **5** The **anterior median fissure** is a deep groove on the anterior surface of the human spinal cord (along the midline). This fissure is analogous to that on the ventral surface of the ox spinal cord.

☐ **6** The **posterior median sulcus** is a somewhat shallower groove along the dorsal surface of the spinal cord.

☐ **7** A **cervical enlargement** is seen in the cervical region of the spinal cord, and a similar **lumbar enlargement** is found in the lower thoracic region. The lumbar enlargement tapers to a cone-shaped end called the **medullary cone.**

☐ **8** The spinal cord proper does not extend below vertebra L1. Instead, separate nerves trail inferiorly until each pair has exited the spinal cavity. This group of nerves, resembling a horse's tail, is called the **cauda equina.**

☐ **9** **Ventral** and **dorsal spinal nerve roots** may be intact in your specimen. These structures are discussed more fully in activity C.

The Spinal Cord

MENINGES
DURA MATER 1
ARACHNOID 2
   SUBARACHNOID SPACE 3
PIA MATER 4

SPINAL CORD
CENTRAL CANAL 5
GRAY COMMISSURE 6
ANTERIOR GRAY HORN 7
POSTERIOR GRAY HORN 8

ANTERIOR WHITE COLUMN 9
POSTERIOR WHITE COLUMN 10
LATERAL WHITE COLUMN 11

SPINAL NERVE
VENTRAL NERVE ROOT 12
DORSAL NERVE ROOT 13
   DORSAL ROOT GANGLION 14
SPINAL NERVE **PROPER** 15
VENTRAL RAMUS 16
DORSAL RAMUS 17

▐◀▌▶ **Figure 19.1** Color the structures and labels of the spinal cord with contrasting colors.

Hint . . . The following structures may or may not be easily seen in the preserved specimen, depending on the specimen's condition. A freshly cut, even cross section usually gives the best results.

□ **10** A cross-sectional view of the spinal cord shows a distinct H-shaped area of gray matter surrounded by areas of white matter. The two lateral sections of gray matter are joined by a transverse **gray commissure.** In the center of the gray commissure, you may see the **central canal.** This canal contains CSF in life but contains preservative in prepared specimens.

□ **11** Each lateral mass of the gray matter exhibits extensions termed the **anterior gray horn** and the **posterior gray horn.** The thoracic and lumbar regions have additional lateral extensions called **lateral gray horns.**

□ **12** Columns of white matter surround the central gray matter area. These include the **anterior white column, lateral white column,** and **posterior white column.** Columns are sometimes called **funiculi.**

B. Microscopic Exploration

SAFETY FIRST!

Observe the usual precautions when using the microscope and slide.

□ **1** Obtain a stained cross-sectional preparation of a mammalian spinal cord and observe it at a low total magnification.

□ **2** Try to find as many features of the spinal cord (as described in activity A) as you can. White matter and gray matter are usually distinguished by differential staining (contrasting colors), as in Minireference Plate 16*d*.

C. Spinal Nerves

Afferent neurons extending through the spinal cord enter as part of a **spinal nerve.** Efferent neurons in the spinal cord exit in spinal nerves. Although they are not really part of the spinal cord proper (they are in the PNS rather than the CNS), we will consider them in this exercise. As each structure is presented, try to locate it on a model or chart of the human nervous system and in figure 19.2.

□ **1** There are 31 pairs of spinal nerves communicating with the spinal cord by way of the intervertebral foramina formed as the vertebrae stack on one another. The spinal nerves are named for the vertebral region with which they are associated:
- ○ **cervical spinal nerves**—C1 through C8 (one pair between the skull and vertebra C1 and one pair inferior to each of seven cervical vertebrae)
- ○ **thoracic spinal nerves**—T1 through T12
- ○ **lumbar spinal nerves**—L1 through L5
- ○ **sacral spinal nerves**—S1 through S5 (branches of these nerves communicate via the dorsal and pelvic foramina of the sacrum)
- ○ **coccygeal spinal nerves**—C or Cx

□ **2** Each spinal nerve communicates with the cord by means of two separate **spinal nerve roots,** each formed by the combination of six to eight tiny *rootlets.* The **ventral (anterior) spinal nerve root** includes efferent fibers (somatomotor and autonomic axons). The **dorsal (posterior) spinal nerve root** contains afferent fibers. Nuclei of afferent neurons form an area of gray matter in the dorsal root called the **dorsal root ganglion.** The spinal nerve roots are found within the spinal cavity. The spinal nerve proper passes through the intervertebral foramina.

□ **3** Outside the spinal cavity, each spinal nerve branches, forming a **ventral ramus** and a **dorsal ramus.** In most of the thoracic and lumbar nerves, the ventral ramus branches into autonomic and somatic pathways. In other segments, the ventral rami's fibers branch to form **plexi** (sing., **plexus**). Plexi are complex nerve networks in which fibers from several different spinal nerves are reorganized to form specific peripheral nerves. Identify each plexus on a model or chart:
- ○ **cervical plexus**
- ○ **brachial plexus**
- ○ **lumbosacral plexus**

Figure 19.2   In the blanks numbered 1 through 31, write the correct spinal nerve abbreviation. In the other blanks, write the name of the cord or nerve feature indicated.

LAB REPORT 19

The Spinal Cord and Spinal Nerves

Name _____

Date _____   Section _____

Figure 19.2 *(from p. 140)*

_____ 1

_____ 2

_____ 3

_____ 4

_____ 5

_____ 6

_____ 7

_____ 8

_____ 9

_____ 10

_____ 11

_____ 12

_____ 13

_____ 14

_____ 15

_____ 16

_____ 17

_____ 18

_____ 19

_____ 20

_____ 21

_____ 22

_____ 23

_____ 24

_____ 25

_____ 26

Summarize your exploration of the spinal cord and spinal nerves by using this checklist.

Structure	Notes
☐ Dura mater	
☐ Arachnoid	
☐ Subarachnoid space	
☐ Pia mater	
☐ Anterior median fissure	
☐ Posterior median sulcus	
☐ Cervical enlargement	
☐ Lumbar enlargement	
☐ Medullary cone	
☐ Cauda equina	
☐ Ventral spinal nerve root	
☐ Dorsal spinal nerve root	
☐ Dorsal root ganglion	
☐ Gray commissure	
☐ Central canal	
Gray horns: ☐ Anterior ☐ Lateral ☐ Posterior	
White columns: ☐ Anterior ☐ Lateral ☐ Posterior	
Spinal nerves: ☐ Cervical ☐ Thoracic ☐ Lumbar ☐ Sacral ☐ Coccygeal	
☐ Ventral ramus of spinal nerve	
☐ Dorsal ramus of spinal nerve	

19-6

Figure 19.2 (continued)

_____ 27

_____ 28

_____ 29

_____ 30

_____ 31

_____ 32

_____ 33

_____ 34

_____ 35

_____ 36

_____ 37

Put in order A

_____ 1

_____ 2

_____ 3

_____ 4

_____ 5

_____ 6

_____ 7

Put in order B

_____ 1

_____ 2

_____ 3

_____ 4

_____ 5

_____ 6

_____ 7

Multiple choice

_____ 1

_____ 2

_____ 3

_____ 4

_____ 5

Put in order A

Put these structures in the order in which a sensory signal might pass.

cervical plexus
dorsal root ganglion
posterior gray horn
receptor
sensory area of the brain
spinal nerve C2
ventral ramus

Put in order B
Put these structures in the order in which a motor signal might pass.

anterior gray horn
effector
lateral gray horn
motor area of the brain
spinal nerve T6
ventral ramus
ventral nerve root

Multiple choice

1. The medial deep groove on the ventral surface of the spinal cord is called the
   a. posterior median sulcus.
   b. medial enlargement.
   c. anterior lateral sulcus.
   d. anterior median fissure.
   e. ventral lateral column.

2. The correct order for the meninges of the brain (from superficial to deep) is
   a. pia mater, dura mater, arachnoid.
   b. dura mater, arachnoid, pia mater.
   c. dura mater, pia mater, arachnoid.
   d. dura mater, subarachnoid, pia mater.
   e. none of the above.

3. The cauda equina is
   a. not present in humans.
   b. a cone-shaped end of the spinal cord proper.
   c. a bundle of separate nerves.
   d. a bundle of cranial nerves.
   e. an enlargement of the spinal cord proper.

4. Spinal nerve branches from several spinal segments (vertical levels) may exchange fibers. Such a network of nerve fibers is called a
   a. braid.
   b. ramus.
   c. plexus.
   d. ganglion.
   e. spinal nerve root.

5. Large groups of neuron cell bodies are likely to be found in
   a. the lateral gray horn.
   b. the gray commissure.
   c. a dorsal root ganglion.
   d. the central canal.
   e. *a, b,* and *c.*

The Brain and Cranial Nerves

Having already explored the spinal cord in Exercise E/19, it is time to move on to an exploration of the other major organ of the central nervous system: the **brain.** The brain occupies most of the cranial cavity within the skull. This rather large, complex organ has been partitioned, for convenience of study, into numerous divisions and subdivisions. Nerves that project directly from the brain are called **cranial nerves.** In this exercise, you are challenged to dissect a mammalian brain (the sheep brain) with cranial nerves and then to apply the knowledge gained to a brief study of human brain and cranial nerve anatomy.

Before You Begin

☐ Read the appropriate chapter in your textbook.

☐ Set your learning goals. When you finish this exercise, you should be able to:
   o list and describe the principal structures of the brain
   o identify important parts of the sheep brain in a preserved specimen
   o identify structures of the human brain in a chart, model, or preserved specimen
   o list the 12 pairs of cranial nerves and describe their general functions

☐ Prepare your materials:
   o dissection tools and trays
   o preserved specimen: *sheep brain (whole)*
   o chart, model, or preserved specimen of a human brain

☐ **Carefully** read the directions and safety tips for this exercise before starting any procedure.

SAFETY FIRST!

Heed the safety instructions that accompany the preservative used with your specimen(s). Avoid cuts and other injuries when using the dissection tools. Consider skin/eye protection.

A. The Sheep Brain: External

The sheep brain is remarkably similar to the human brain. One major difference is in proportion. For example, the

sheep has a proportionally smaller **cerebrum.** Another difference is in orientation. The sheep brain is oriented anterior-to-posterior, as in any four-legged animal. The human brain is oriented superior-to-inferior. Examine the external features of the sheep brain as outlined in these steps:

☐ **1** Determine whether the **meninges** of your specimen are still intact. If so, find the following:
   o **dura mater**—a tough, leathery outer covering (the dura forms a large cavity for blood along the top of the brain called the **sagittal sinus**)
   o **arachnoid**—the thin, translucent middle meninx (recall that the *subarachnoid space* contains cerebrospinal fluid, CSF, in living specimens)
   o **pia mater**—a thin, vascular membrane covering the brain (preserved blood in the vessels may give it a dark brown appearance)

☐ **2** The most prominent feature of the brain is the **cerebrum** (fig. 20.1). This region is divided into nearly symmetrical left and right **hemispheres** by a deep **longitudinal fissure.** It is along this fissure that the sagittal sinus of the dura lies. Bend the two hemispheres away from each other and observe the **corpus callosum,** a mass of white fibers that connects the two hemispheres. The surface of the cerebrum is covered with large folds of tissue called **gyri** (sing., **gyrus**). The grooves between the gyri are **sulci** (sing., **sulcus**). The deeper sulci are often termed **fissures.** Anatomists use the fissures as landmarks to divide the surface of the cerebrum (the **cerebral cortex**) into regions that roughly correspond to the overlying skull bones:
   o **frontal lobes**
   o **parietal lobes**
   o **occipital lobes**
   o **temporal lobes**
   o **insula** (buried within a fissure between the temporal and parietal lobes)

☐ **3** The smaller, rounded structure caudal to the cerebrum (and dorsal to the brainstem) but still possessing gyri is the **cerebellum.** The cerebellum has smaller gyri that are roughly parallel to one another, unlike the convoluted gyri of the cerebrum.

☐ **4** The **diencephalon** is the brain region ventral to the cerebrum and cranial to the brainstem. One can see some of its features in a ventral view of the sheep brain. The **hypothalamus** includes the **optic chiasma,** an X-shaped junction of fibers at the bases of the **optic nerves.** The legs of the chiasma attached to the brain are the beginnings of the **optic tracts.** The **mammillary bodies,** also part of the hypothalamus, are just posterior to the optic chiasma. The stalklike **infundibulum** projects ventrally from between the chiasma and the mammillary bodies. The infundibulum connects the hypothalamus to the **pituitary gland,** or *hypophysis.* Another major part of the diencephalon is the **thalamus,** which is not visible in an external examination.

Hint . . . If your specimen has an intact pituitary (not all do), it will be in a mass of tissue on the ventral aspect. To see the infundibulum and adjacent structures, it is necessary to pull the pituitary away from the ventral surface of the brain. You may be able to keep the infundibulum in place as you do the separation, as if it were a hinge on a flap of tissue. Some cranial nerves may be pulled away from the brain, so cut them in a way that leaves as much of the nerve attached to the brain as possible.

☐ **5** Holding the specimen in gloved hands, dorsal aspect toward you, carefully bend it to expose the tissue between the cerebrum and cerebellum (fig. 20.2). You may have to tease some connecting tissue apart with a probe. You now see the dorsal surface of the **midbrain.** A small, rounded body on the midline, nearest the cerebrum, is the **pineal body** of the diencephalon. The four rounded bumps caudal to the pineal body constitute the **corpora quadrigemina** of the midbrain. The four bumps are distinguished as the **superior colliculi** (the slightly larger pair) and **inferior colliculi** in the human. The midbrain is part of the brain division called the **brainstem,** which is found between the diencephalon and the spinal cord.

☐ **6** The portion of the brainstem just caudal to the midbrain is the **pons.** The cerebellum covers the dorsal surface of the pons, but the ventral surface of the pons can be clearly seen as a large bulge.

☐ **7** The **medulla,** or *medulla oblongata,* is caudal to the pons. The medulla looks as if it were a swollen region of the spinal cord, to which it is cranial. Together, the midbrain, pons, and medulla constitute the brainstem.

☐ **8** Because of their smallness and the manner in which specimens are usually prepared, some of the **cranial nerves** may be impossible to locate on your specimen. Normally, there are 12 pairs of nerves extending from various parts of the brain. These nerves, of course, are part of the peripheral nervous system (PNS) rather than part of the brain proper. Using figure 20.1*a* as a guide, try to locate as many cranial nerves as you can.

B. The Sheep Brain: Internal

☐ **1** Use a knife or long-bladed scalpel to cut the specimen along a median (midsagittal) plane. Use the longitudinal fissure as a cutting guide (fig. 20.3 and 20.4).

☐ **2** The corpus callosum had been connecting the two cerebral hemispheres and can now be clearly seen in the brain section. The **fornix,** part of the diencephalon, is ventral to the corpus callosum and is also composed of white fibers. You may be able to see a hollow cavity just ventral to the corpus callosum in each brain half. These cavities are the right and left **lateral ventricles.** In the whole brain, they are separated by a thin membrane, the **septum pellucidum.** Because of variations in the plane of sectioning, the septum may be visible as a drumlike covering to one ventricle or may partially cover both ventricles. The **choroid plexus,** a tiny mass of capillaries within each ventricle, produces the cerebrospinal fluid that fills each ventricle.

☐ **3** The lateral ventricles can also be seen in a frontal section. Use one of the sheep brain halves or a new whole brain to cut a frontal section. The infundibulum is a good landmark for the correct position of the frontal cut. Locate the lateral ventricle(s). Just lateral to each lateral ventricle is a small circle of gray matter called the **caudate nucleus,** part of the cerebrum's *basal ganglia.* Just inferior to the lateral ventricles, along the midline, is a single **third ventricle.** Connecting each lateral ventricle to the third ventricle is an **interventricular foramen (of Monro),** which is not visible in any of the brain sections. The lateral walls of the third ventricle are formed by a mass of gray matter called the **thalamus,** a region of the diencephalon. The two *lateral masses* of the thalamus are joined by the *intermediate mass* that passes through the third ventricle. Take note of the distribution of gray and white matter in the frontal section. What do you notice about its pattern?

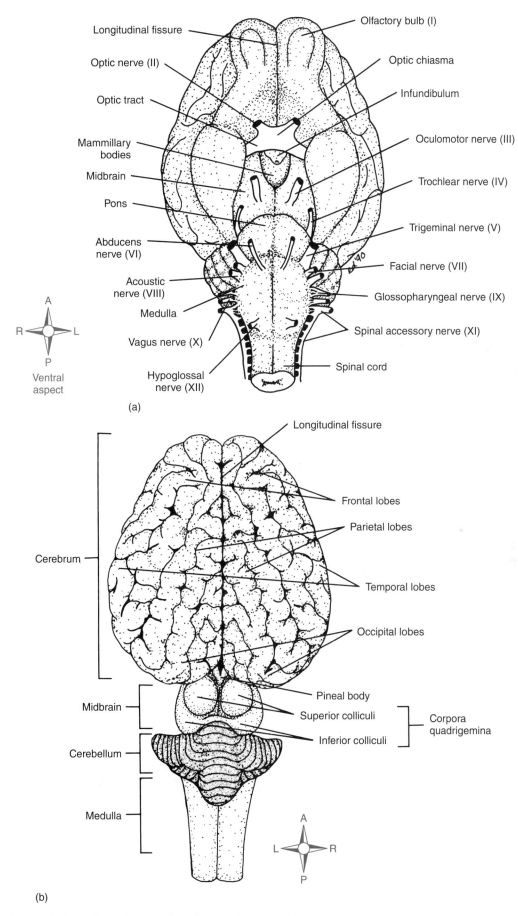

Figure 20.1 External view of the sheep brain: (a) ventral aspect, (b) dorsal aspect.

□ **4** Return your attention to the median (midsagittal) section. Identify the white matter pattern, or **arbor vitae,** of the cerebellum. Ventral to the cerebellum is the **fourth ventricle,** which is connected to the third ventricle by the **cerebral aqueduct** (aqueduct of Sylvius).

□ **5** Many features of the diencephalon already seen on the external aspect or in the frontal section can be seen in the median (midsagittal) section. Locate the pineal body, optic chiasma, mammillary bodies, and infundibulum. In this section, only the intermediate mass of the thalamus is visible. It appears as a circle of gray matter surrounded by the shallow section of the third ventricle.

□ **6** The corpora quadrigemina is easily seen in the median (midsagittal) section, along the dorsal surface of the midbrain, just posterior to the pineal body. The white matter tracts in the ventral portion of the midbrain constitute the **cerebral peduncles.**

□ **7** Identify the pons and medulla in the median (midsagittal) section see figures 20.3 and 20.4.

C. The Human Brain

Now that you have explored mammalian brain anatomy by reading the appropriate section of your textbook and by studying the sheep brain specimen, you are ready to apply your knowledge to a study of the human brain (figs. 20.5 and 20.6).

Using a dissectible model of the human brain, a chart, or a preserved human brain specimen, identify all of the structures that you did for the sheep brain. Use the checklist in Lab Report 20 as a guide.

D. Cranial Nerves

Cranial nerves are nerves that project from the brain, mainly the brainstem. There are 12 pairs of cranial nerves. Each pair is known both by a Roman numeral and by a name. Cranial nerves are numbered in order according to their origin from the ventral surface of the brain. The low-numbered cranial nerves are anterior; the high-numbered nerves are posterior, as you can see in figure 20.7.

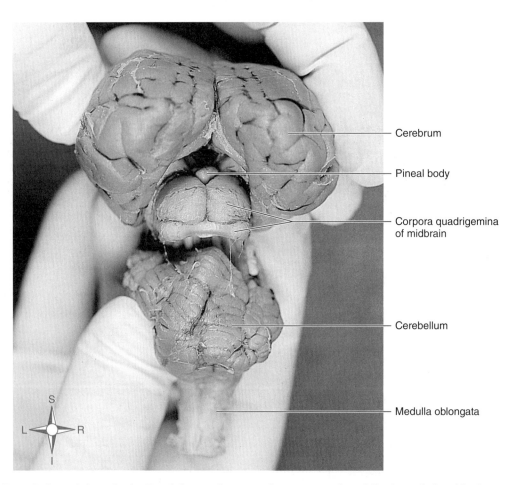

Figure 20.2 Dorsal view of sheep brain. Bend the specimen gently to expose the midbrain and pineal body.

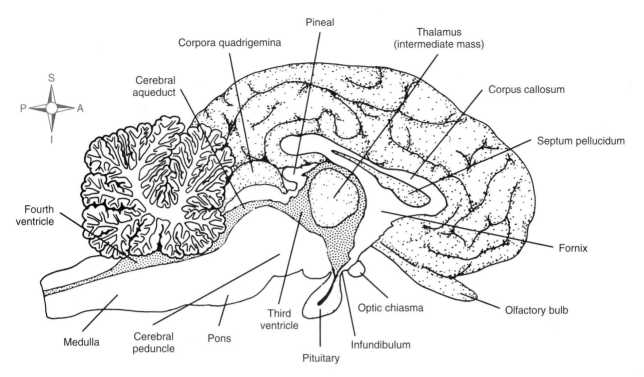

Figure 20.3 Midsagittal section of the sheep brain.

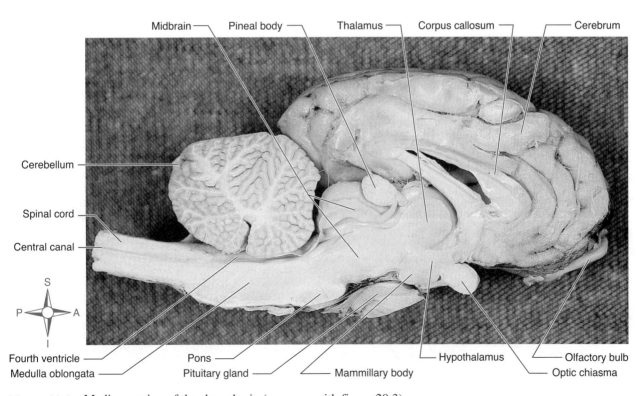

Figure 20.4 Median section of the sheep brain (compare with figure 20.3).

The Human Brain

CEREBRUM

CORPUS CALLOSUM 1
LATERAL VENTRICLE 2

CEREBELLUM

GRAY MATTER 3
ARBOR VITAE 4
FOURTH VENTRICLE 5

DIENCEPHALON

OPTIC CHIASMA 6
INFUNDIBULUM 7
MAMMILLARY BODIES 8

THALAMUS **INTERMEDIATE MASS** 9
PINEAL BODY 10
FORNIX 11
THIRD VENTRICLE 12

MIDBRAIN

CEREBRAL PEDUNCLES 13
CEREBRAL AQUEDUCT 14
CORPORA QUADRIGEMINA 15
PONS 16
MEDULLA 17

Figure 20.5   Use pens or pencils of contrasting colors to shade in the structures of the human brain.

Some cranial nerves contain only sensory (afferent) fibers. Some cranial nerves contain primarily somatic motor (efferent) fibers. These "motor nerves" also have a few sensory fibers for feedback on muscle tension. Some cranial nerves have significant numbers of both sensory fibers and motor fibers—and sometimes also parasympathic (efferent) nerves of the autonomic nervous system. Table 20.1 clarifies these general functions.

Locate each of the 12 pairs of cranial nerves listed in table 20.1 and figure 20.7 in a model or specimen.

Hint . . . The names and numbers of the cranial nerves can be remembered easily by using a mnemonic aid, such as a sentence or phrase with words that start with the same letters as the cranial nerves, in the correct order—for example, *On old Olympus's tiny tops, a friendly Viking grew vines and hops.*

Make up your own sentence. The more outlandish and comical, the more likely you will remember it. Of course, you can also use this hint for other lists you have to memorize.

TABLE 20.1   *Cranial Nerves*

| Cranial Nerve | | General Functions | | |
Number	Name	Sensory	Somatic Motor	Parasympathetic
I	Olfactory	Smell		
II	Optic	Vision		
III	Oculomotor		Upper eyelid (blinking); four (of six) extrinsic eye muscles (eye movement)	Pupil constriction; thickening of lens
IV	Trochlear		Extrinsic eye muscle, superior oblique (helps depress eye)	
V	Trigeminal	Sensation in face, teeth	Muscles of mastication (chewing)	
VI	Abducens		Extrinsic eye muscle, lateral rectus (abducts eye)	
VII	Facial	Taste	Facial expression	Salivation; production of tears
VIII	Vestibulocochlear	Hearing, equilibrium		
IX	Glossopharyngeal	Taste, touch	Pharyngeal muscles (swallowing)	Salivation
X	Vagus	Sensation in pharynx, larynx, viscera	Muscles of palate, pharynx, larynx (swallowing, voice production)	Regulation of viscera of thorax and abdomen
XI	Accessory		Two muscles of neck and upper back (helps in moving head, neck, shoulders)	
XII	Hypoglossal		Tongue muscles (swallowing, speaking, sticking your tongue out at your lab partner)	

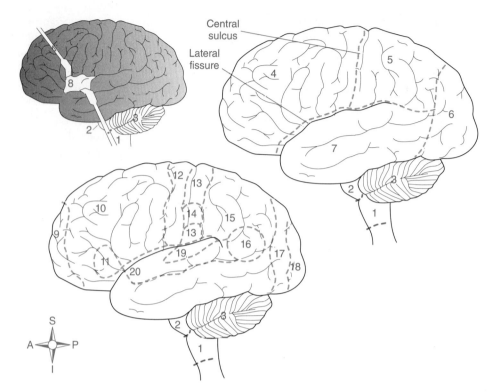

Central sulcus

Lateral fissure

S
A — P
I

Figure 20.6   Use contrasting colored pens or pencils to color these figures of the external surface of the brain.

Brainstem
MEDULLA OBLONGATA, PONS 2
MIDBRAIN (NOT VISIBLE)
   CEREBELLUM 3

Cerebrum: Lobes of Cortex
FRONTAL LOBE 4
PARIETAL LOBE 5
OCCIPITAL LOBE 6
TEMPORAL LOBE 7
INSULA 8

Cerebrum: Functional Areas of Cortex
PREFRONTAL AREA 9

PREMOTOR AREA 10
MOTOR SPEECH AREA 11
PRIMARY SOMATIC MOTOR AREA 12
PRIMARY SOMATIC SENSORY AREA 13
PRIMARY TASTE AREA 14
SOMATIC SENSORY ASSOCIATION AREA 15
SENSORY SPEECH AREA 16
VISUAL ASSOCIATION AREA 17
VISUAL CORTEX 18
PRIMARY AUDITORY AREA 19
AUDITORY ASSOCIATION AREA 20

Figure 20.7 Use contrasing colored pens or pencils to color this figure showing the ventral surface of the brain and the cranial nerves.

Cranial Nerves

I   OLFACTORY NERVE   1

II   OPTIC NERVE   2

III   OCULOMOTOR   3

IV   TROCHLEAR NERVE   4

V   TRIGEMINAL NERVE   5

VI   ABDUCENS NERVE   6

VII   FACIAL NERVE   7

VIII   VESTIBULOCOCHLEAR NERVE   8

IX   GLOSSOPHARYNGEAL NERVE   9

X   VAGUS NERVE   10

XI   ACCESSORY NERVE   11

XII   HYPOGLOSSAL NERVE   12

Magnetic Resonance Imaging

As you recall, traditional radiographic imaging relies on x-rays that easily pass through soft tissues but cast shadows when they try to pass through dense tissues. Thus, soft tissues can be seen only when they are coated or filled either with a substance that blocks radiation or with a substance that emits its own radiation.

A newer technology that produces finely detailed images of soft tissue without relying on x-rays is **nuclear magnetic resonance (NMR).** Called **magnetic resonance imaging (MRI)** in many clinical settings, this technique relies on the magnetic properties of the nuclei of hydrogen molecules in human tissues.

When placed in a strong magnetic field, the hydrogen nuclei in a living tissue all orient themselves in one direction. At different intervals, depending on the type of tissue, the nuclei spin back to their original position and give off a burst of RF (radio frequency) energy. A computer with sensors that can detect the relative position and rate of energy bursts then constructs a three-dimensional image of the tissues. Thus, a subject is placed in a magnetic field, and the MRI monitor displays sections similar to, but clearer than, CT images.

The brain is a particularly interesting subject for MRI because it is composed entirely of soft tissue. Although CT scanning technology can produce images of brain structure, MRI pictures are often much more detailed. The fact that patients scanned with MRI receive no harmful nuclear radiation makes this technology even more attractive.

Label the brain parts indicated in the following magnetic resonance image.

The Brain
and Cranial Nerves

Name _____

Date _____ Section _____

Use this table as a checklist to keep track of your study of the sheep brain specimen and human model.

Brain Region	Structure	Sheep	Human	Notes
Meninges	Dura mater	☐	☐	
	Arachnoid	☐	☐	
	Pia mater	☐	☐	
Cerebrum	Hemispheres	☐	☐	
	Longitudinal fissure	☐	☐	
	Corpus callosum	☐	☐	
	Frontal lobe	☐	☐	
	Parietal lobe	☐	☐	
	Occipital lobe	☐	☐	
	Temporal lobe	☐	☐	
	Insula	☐	☐	
	Lateral ventricle	☐	☐	
	Septum pellucidum	☐	☐	
	Caudate nucleus	☐	☐	
Cerebellum	Gray matter	☐	☐	
	Arbor vitae	☐	☐	
	Fourth ventricle	☐	☐	
Diencephalon	Optic chiasma	☐	☐	
	Infundibulum	☐	☐	
	Mammillary bodies	☐	☐	
	Thalamus	☐	☐	
	Pineal body	☐	☐	
	Fornix	☐	☐	
	Third ventricle	☐	☐	
Midbrain	Cerebral peduncles	☐	☐	
	Cerebral aqueduct	☐	☐	
	Corpora quadrigemina	☐	☐	
Pons	Pons	☐	☐	
Medulla	Medulla	☐	☐	

20-12

MRI human brain (*from p. 152*)

_____ 1

_____ 2

_____ 3

_____ 4

_____ 5

_____ 6

_____ 7

_____ 8

_____ 9

Identify

_____ 1

_____ 2

_____ 3

_____ 4

_____ 5

_____ 6

_____ 7

_____ 8

_____ 9

_____ 10

_____ 11

_____ 12

_____ 13

_____ 14

_____ 15

_____ 16

_____ 17

_____ 18

_____ 19

_____ 20

Put in order

_____ 1

_____ 2

_____ 3

_____ 4

_____ 5

Identify

1. A mass of white fibers connecting the left and right cerebral hemispheres
2. A cerebral lobe not visible in an ordinary external inspection
3. A fold of cortical gray matter on the surface of the cerebrum
4. A deep sulcus
5. A vein in the dura mater roughly parallel to the longitudinal fissure
6. A rounded structure dorsal to the brainstem
7. A distinct, branched pattern of white matter in the cerebellum
8. The middle of three divisions of the brainstem
9. A vascular structure that produces CSF and is present in all the fluid ventricles
10. The fluid ventricle associated with the cerebellum
11. The membrane that separates the left and right lateral ventricles
12. The portion of the thalamus that passes through the third ventricle
13. The structure composed of four colliculi
14. The gland that is not technically a part of the brain but is attached to the brain via the infundibulum
15. The gray matter on the surface of the cerebrum
16. The white matter tracts in the ventral portion of the midbrain
17. The most caudal structure of the brain
18. The cranial nerve that regulates most of the internal thoracic and abdominal organs
19. The cranial nerve that carries vision information from the eye to the brain
20. The cranial nerve that helps you hear and keep your balance

Put in order

Arrange these in the order in which CSF passes through them, beginning with the place in which CSF could first be formed.

cerebral aqueduct

interventricular foramen (of Monro)

fourth ventricle

lateral ventricle

third ventricle

The Eye and Vision

Vision, one of the *special senses,* relies on a very complex receptor apparatus, the **eye.** Vision relies on a variety of positioning and focusing mechanisms to form an image in the correct spot on the light-sensitive receptor cells inside the eye. These mechanisms involve muscles, lenses, and other structures that are all a part of the visual apparatus. The complexity of structure allows for complexity of function. The visual image perceived by humans has the qualities of **resolution, brightness, color,** and **depth.** These and other aspects of visual structure and function are presented in this exercise.

Before You Begin

☐ Read the appropriate chapter in your textbook.

☐ Set your learning goals. When you finish this exercise, you should be able to:
  o describe the major features of the eye and identify them in specimens, models, and figures
  o explain the basic function of the eye and its structures
  o demonstrate visual function by means of standard tests and demonstrations

☐ Prepare your materials:
  o sheep or beef eye (fresh or preserved)
  o dissection tools and trays
  o chart or dissectible model of the eye
  o meter stick
  o wall chart: *astigmatism test*
  o wall chart: *Snellen eye test*
  o pack of assorted brightly colored papers
  o color-blindness test plates (Ichikawa Test Book or similar)

☐ **Carefully** read the directions and safety tips for this exercise before starting any procedure.

A. The Human Eye

The human eye and its accessory structures provide the structural apparatus required for processing and receiving visual images. Using charts or dissectible models of the

human eye and its accessories, locate each of the parts listed and describe its function in Lab Report 21.

☐ **1** Locate these features of the **lacrimal apparatus** (fig. 21.1):
  o **lacrimal gland**—the exocrine tear gland in the superior lateral corner of the orbit
  o **nasolacrimal duct**—the duct in the inferior medial corner of the orbit; drains tears from **lacrimal canaliculi** toward the nasal cavity

☐ **2** Locate these *extrinsic eye muscles,* which are skeletal muscles that control eye movement, and determine what the action is of each of these muscles:
  o **rectus muscles**—(four) *superior, inferior, medial, lateral*
  o **oblique muscles**—(two) *superior, inferior*

☐ **3** Identify these additional accessory structures:
  o **eyelids**
  o **conjunctiva**—the thin, transparent mucous membrane adhering to the anterior surface of the eye and lining the eyelids
  o **eyebrows**

☐ **4** Distinguish among the three **tunics,** or *coats,* of the eyeball:
  o **Fibrous tunic**—the *outer coat* of the eye. The cornea is the anterior, transparent portion. The sclera is the white, fibrous portion.

Figure 21.1 Label these parts of the lacrimal apparatus.

○ **Vascular tunic**—the *middle coat* of the eye. The posterior portion is the thin, heavily pigmented choroid. The anterior portions include the circular ciliary body, in which are incorporated the **ciliary muscles** (*intrinsic eye muscles*). The ciliary muscles are attached to the rim of the lens by means of **suspensory ligaments.** Attached to the anterior edge of the ciliary body is the colored **iris,** which has an opening called the pupil.

○ **Nervous tunic**—the *inner coat* of the eye. The retina is divided into two layers: the outer *pigmented retina* and the inner *sensory retina.* The sensory retina contains photoreceptor cells of two types, **rods** and **cones,** as well as association neurons. In the posterior retina is a small, yellow **macula** (*macula lutea*) with a pit called the **fovea centralis.** The fovea is normally the center of the visual field and contains many cones. Rods become more dominant farther away from the fovea. Medial to the macula lutea is the white **optic disc.** Here, blood vessels, as well as the nerve fibers that exit as the **optic nerve,** pass out of the eyeball.

☐ **5** Identify these compartments of the eye:
○ **Anterior compartment**—the *anterior cavity,* filled with aqueous humor, a watery filtrate produced by the ciliary body. Aqueous humor circulates from the **posterior chamber** behind the iris, through the pupil, to the **anterior chamber,** where it is reabsorbed.
○ **Posterior compartment**—the *posterior cavity,* filled with transparent, jellylike **vitreous humor.** This cavity is posterior to the lens.

B. Eye Dissection

Apply your knowledge of mammalian eye anatomy by carefully dissecting a beef or sheep eye (fig. 21.2).

SAFETY FIRST!

Observe the usual safety precautions while dissecting the preserved or fresh tissue of the eye specimen.

☐ **1** Trim away any excess adipose tissue, leaving the stub of the optic nerve and extrinsic eye muscles intact.
☐ **2** Locate these structures on the external aspect:
○ **optic nerve**—the nerve bundle projecting from the posterior of the eyeball
○ **extrinsic eye muscles** (six)—these may have been cut from your specimen during preparation
○ **sclera**—the white portion of the fibrous tunic

○ **cornea**—the clear, anterior portion of the fibrous tunic
○ **iris**—the pigmented region under the cornea
○ **pupil**—the iris's hole (it may be oblong in your specimen rather than round as in the human)
○ **conjunctiva**—the thin mucous membrane over the anterior portion of the eye, extending to line the inner eyelid
○ any other accessory structures that may be present in your specimen

☐ **3** Puncture the sclera with the tip of your scissors (or a scalpel) about 1 cm posterior to the edge of the cornea. Use the scissors to cut a circle around the eye, staying 1 cm from the cornea's margin. Pull the anterior portion away from the posterior portion.
☐ **4** Identify these features:
○ **vitreous humor**—the thick fluid of the posterior compartment
○ **aqueous humor**—a watery liquid in the anterior compartment
○ **lens**—a plasticlike ball of translucent tissue (in fresh specimens it is very clear; hold it up and try to look through it)
○ **ciliary body**—a ring of ridges around the outside of the iris's margin (the ridges are formed by the **ciliary muscles** within)
○ **retina**—a thin film of gray matter loosely associated with the inside posterior wall of the eyeball (it may have fallen away, appearing as a crumpled mass still attached at the optic disc)
○ **choroid**—the pigmented region of the vascular tunic in the posterior region of the eyeball's wall (deep to the sclera, superficial to the retina) (in some mammals but not in humans, an iridescent **tapetum lucidum** can be seen in the choroid)

C. Visual Function

Human vision is a complex function involving **image formation, reception** by photoreceptors, **transmission** of sensory information to appropriate brain centers, and **perception** of the image in the mind. The activities described in this exercise challenge you to demonstrate just a few of the many functional characteristics of the human sense of vision.

☐ **1** **Acuity** is the sharpness of the visual image and is often tested in clinics with the *Snellen eye test.*
○ Place a Snellen chart on the wall and have the subject stand 20 feet (6.1 meters) away, facing the chart.
○ Have the subject cover one eye and read the letters on the chart as you point to them. Begin at the top and work your way down.

The Eye

Figure 21.2 Color the labels and the parts of the eye and its accessory structures with contrasting colors.

o Note the lowest row that can be read accurately. Record the number printed next to that row in Lab Report 21. That number is the farthest distance (in feet) that a person with normal acuity can see the letters in that row. For example, if the number is 20, the subject has 20-20 vision, meaning the subject can see at 20 feet what a person with normal vision can see at 20 feet. If the number is 100, the subject has 20-100 vision, meaning the subject sees at 20 feet what a person with normal vision can see at 100 feet. Test both eyes; if the subject wears eyeglasses, try the test corrected (with glasses) and uncorrected (without glasses).

□ 2 **Astigmatism** is abnormal curvature of the cornea.
   o Test for a curvature flaw by facing the subject toward an *astigmatism test chart* 20 feet (6.1 meters) away.
   o Cover one of the subject's eyes and ask him or her to stare at the circle in the center of the chart.
   o If all the lines radiating from the center appear to the subject to be straight and of equal darkness, no astigmatism is present. If the lines appear wavy, curved, or of unequal darkness, astigmatism is suspected. Test the other eye.

□ 3 **Accommodation** is the ability to adjust the focusing apparatus to account for changes in distance from the viewed object. Your lens changes shape to accommodate near vision. Likewise, your pupil changes diameter to help accommodation.
   o Position the subject's chin on the lab table and put a meter stick on the table perpendicular to the subject's face (with 0 cm at the chin).
   o Have the subject cover one eye and ask him or her to focus on the word *TEST* in figure 21.3 as you hold it 1 meter from the face.
   o Slowly move the figure closer and record the nearest point at which the subject can focus. Does the diameter of the pupil change as you approach the subject's *near point?* Repeat with the other eye.

TEST

Figure 21.3 Near-point accommodation test figure.

□ 4 The **blind spot** is actually the optic disc, a point where there are no photoreceptors in the retina. The part of the *visual field* focused on the blind spot cannot be seen. We normally don't notice it because the brain "fills in" the blank area for us. Demonstrate the location of the blind spot in your visual field by covering one eye and looking at figure 21.4. Stare at the object (square or circle) *medial* to the other object in the figure. Starting at about 35 cm, slowly bring the figure closer to your eye. At one point, the *lateral* object will seem to disappear because its image has fallen on the blind spot.

Figure 21.4 The blind spot test objects.

□ 5 Rods are photoreceptors adapted to reception of dim light of a broad color range (that is, *dim light–black-and-white vision*). Cones are adapted for *bright light–color vision* because they require higher light intensity, but different cones can perceive different wavelengths (colors) of light. The cones are densest at the fovea. The rods are densest in the peripheral portion of the visual field. You can demonstrate the *monochrome* nature of the peripheral field:
   o While the subject stares forward, slowly bring a brightly colored sheet of paper into the visual field from behind the subject's head.
   o Stop when the subject indicates that the object has just entered the visual field.
   o Ask what color the object is. If you have done the test properly and the subject isn't cheating, the color is difficult or impossible to determine. Why?

□ 6 **Color blindness,** or color vision deficiency, can be the result of brain or nerve damage but is usually an inherited condition. In the more common forms, *red-green deficiency,* red- or green-sensitive cones are missing or defective. Individuals with no green-sensitive cones have *deuteranopia;* those with no red-sensitive cones have *protanopia.* Test for red-green deficiency in your subject by following these steps (for the Ichikawa Test Book) or the steps with your test kit:
   o Hold the book of plates about 75 cm from the subject in bright, natural light, avoiding glare.
   o *Without touching the figures,* ask the subject to name the numeral seen in each mosaic. If more than one numeral can be seen, ask for the one that appears most distinct. Record the numeral for each plate on the form provided with the test kit.
   o Add the marks in the two columns of the form's first section. Using the formula given there, determine whether the subject is normal or color deficient.
   o *Only if the subject is color deficient,* complete the second section of the form. Determine whether the subject has protanopia or deuteranopia.

LAB REPORT 21

The Eye and Vision

Name _____

Date _____ Section _____

Use this table as a checklist for your study of the human and sheep (or beef) eye. Don't forget to fill in the function(s) column.

Tunic or Region	Structure	Sheep	Human	Function(s)
Lacrimal apparatus	Lacrimal gland	☐	☐	
	Nasolacrimal duct	☐	☐	
Extrinsic eye muscles	Rectus: superior, inferior, lateral, medial	☐	☐	
	Oblique: superior, inferior	☐	☐	
External accessories	Conjunctiva	☐	☐	
	Eyelids, eyebrows	☐	☐	
Fibrous tunic	Cornea	☐	☐	
	Sclera	☐	☐	
Vascular tunic	Choroid	☐	☐	
	Ciliary body	☐	☐	
	Lens	☐	☐	
	Suspensory ligaments	☐	☐	
	Iris	☐	☐	
	Pupil	☐	☐	
	Tapetum lucidum	☐	☐	
Nervous tunic	Retina	☐	☐	
	Macula	☐	☐	
	Fovea	☐	☐	
	Optic disc	☐	☐	
Anterior compartment	Aqueous humor	☐	☐	
Posterior compartment	Vitreous humor	☐	☐	
	Optic nerve	☐	☐	

A second data table is on the next page.

Test or Demonstration	Results		Remarks
	Left Eye	Right Eye	
Snellen (acuity) test	Corrected: 20-_____	Corrected: 20-_____	
	Uncorrected: 20-_____	Uncorrected: 20-_____	
Astigmatism	Corrected: ☐ Normal ☐ Not normal	Corrected: ☐ Normal ☐ Not normal	
	Uncorrected: ☐ Normal ☐ Not normal	Uncorrected: ☐ Normal ☐ Not normal	
Accommodation (near point)	Corrected: _____ cm	Corrected: _____ cm	
	Uncorrected: _____ cm	Uncorrected: _____ cm	
Blind spot	☐ Found	☐ Found	
Peripheral color perception	☐ Correct identification ☐ Incorrect identification	☐ Correct identification ☐ Incorrect identification	
Color-blindness screening	☐ Normal ☐ Protanopic ☐ Deuteranopic	☐ Normal ☐ Protanopic ☐ Deuteranopic	

Figure 21.1

_____ 1

_____ 2

_____ 3

Fill-in

_____ 1

_____ 2

_____ 3

_____ 4

_____ 5

_____ 6

_____ 7

_____ 8

Put in order

_____ 1

_____ 2

_____ 3

_____ 4

_____ 5

_____ 6

Fill-in

1.   __?__ is a type of photoreceptor specialized for vision in dimly lit environments.
2.   Abnormal curvature of the cornea may produce a condition called __?__.
3.   Tears are produced by the __?__ glands.
4.   When you try to keep your eyes focused on a rocket as it ascends, you contract the __?__ muscle attached to your eyeballs.
5.   Muscles in the __?__ relax or contract to alter the diameter of the pupil.
6.   The relative sharpness of an image formed in your eyes is called your __?__.
7.   The redness seen in one's eyes in a smoke-filled environment is a result of temporary inflammation of the __?__.
8.   On a hot day, sweat collects on your forehead. It is prevented from dripping directly toward your eye by the __?__.

Put in order

Arrange these in the order in which light passes through them to form an image inside the eye.

conjunctiva

cornea

lens

pupil

retina

vitreous humor

The Ear, Hearing, and Equilibrium

Hearing, like vision, is considered to be one of the *special senses.* Hearing, the ability to detect a range of sound **frequencies** (pitches) and **intensities,** is only one of the senses mediated by the ear apparatus. Two types of **equilibrium** are also mediated by ear receptors. One type, **static equilibrium,** allows you to determine your position relative to a center of gravity. In other words, it tells you whether you're right side up, upside down, or something in between. Your sense of **kinetic (dynamic) equilibrium** gives information regarding the speed and direction of your body's motion.

This exercise challenges you to learn the basic divisions and structures of the human ear, as well as how they function in hearing and equilibrium.

Before You Begin

☐ Read the appropriate chapter in your textbook.

☐ Set your learning goals. When you finish this exercise, you should be able to:
   o distinguish among the three main divisions of the ear
   o identify the principal structures of the human ear in models and figures
   o explain how ear structures sense sound and equilibrium stimuli
   o demonstrate some clinical screening tests for hearing and equilibrium

☐ Prepare your materials:
   o model or chart of the human ear
   o microscope
   o prepared microslide: *cochlea c.s.*
   o tuning fork (256 Hz)
   o swimmer's ear plugs (optional)
   o chalkboard and chalk
   o bright desk lamp
   o metric ruler (30 to 100 cm)
   o watch or clock

☐ **Carefully** read the directions and safety tips for this exercise before starting any procedure.

A. Ear Structure and Function

Use a chart or model of the human ear to locate and study the ear structures listed. You may want to use your textbook as an additional aid in this activity.

☐ **1** The **external (outer) ear** is the ear division composed of the following elements:
   o **Auricle (pinna)**—This external ear flap protects the auditory opening, directing sound waves toward it. It also functions as a "radiator" in thermoregulation.
   o **External auditory canal** (also known as the *External acoustic meatus*)—This tubelike passage carries airborne sound waves farther into the ear apparatus.
   o **Tympanic membrane**—Also called the *eardrum,* it covers the end of the external auditory canal to form a boundary with the middle ear. It vibrates when struck by airborne sound waves, carrying the sound energy into the middle ear.

☐ **2** The **middle ear** begins where the external ear ends, with the tympanic membrane. The middle ear is an air-filled cavity lined with mucous membrane. Identify these important parts of the middle ear:
   o **Malleus**—It is one of three *auditory ossicles* in each ear. Also called the *hammer,* it is a tiny, club-shaped bone attached to the eardrum. It vibrates when sound waves pass to it from the eardrum.
   o **Incus**—Also called the *anvil,* this tiny bone forms a synovial joint with the malleus. The incus vibrates when it receives energy from the malleus.
   o **Stapes**—Called the *stirrup* because of its shape, this ossicle is joined to the incus, from which it receives vibrations. A flat portion of the stapes fits into the **oval window,** a passage into the inner ear. Because of the structural relationship of this chain of ossicles, sound waves are carried from the tympanic membrane to the oval window (that is, from the external ear to the inner ear). Along the way, the vibrations are amplified for better reception.
   o **Auditory (eustachian) tube**—It is a collapsible tube running between the middle ear and the

pharynx. It allows internal air pressure to equalize with atmospheric air pressure so that high pressure on one side does not distort or muffle the eardrum.

o The middle ear also has an opening to the mastoid air cells in the mastoid process of the temporal bone. What significance might this have if the middle ear becomes infected?

SAFETY FIRST!

Observe the usual safety precautions when using the microscope and prepared slides.

☐ **3** The **inner ear** is the third division of the ear apparatus. It is here that receptors for hearing and equilibrium are located. The inner ear is within a hollow area in the petrous portion of the temporal bone. A mazelike **bony labyrinth** contains a similarly shaped but smaller **membranous labyrinth.** The fluid inside the membranous labyrinth is called **endolymph,** and the fluid outside the membranous labyrinth is called **perilymph.** The bony labyrinth, and the membranous labyrinth inside it, is composed of three main regions:

o **Cochlea**—It is a long passage coiled like a snail. A cross section reveals that the passage is divided into three chambers by a Y-shaped partition. The base of the Y is a projection of the bone called the **spiral lamina.** The two branches are pieces of membrane called the **vestibular membrane** and the **basilar membrane.** The space between the two membranes is the endolymph-filled **cochlear duct.** The space outside the vestibular membrane is the **scala**

vestibuli, whereas the **scala tympani** is the space outside the basilar membrane. Sound waves move into the perilymph of the scala vestibuli as the stapes vibrates in the oval window. The vestibular membrane vibrates, causing the endolymph in the cochlear duct to vibrate. This, in turn, causes the flaplike **tectorial membrane** to vibrate and bend the hairs projecting from the **spiral organ** or **organ of Corti** on the basilar membrane (which also vibrates). The bending of hairs induces receptor potentials in the sensory neurons. The energy dissipates as it moves through the scala tympani to the round window. A branch of cranial nerve VIII called the **cochlear nerve** carries information to auditory areas in the brain. Observe a cross section of a mammalian cochlea in the microscope and sketch your observations. Label as many parts as you can identify.

o **Vestibule**—This is the central area of the inner ear. The vestibule contains saclike portions of the membranous labyrinth called the **utricle** and the **saccule.** Each has a patch of sensory hair cells called a **macula** (fig. 22.1a). The macula's hair cells are covered by a gelatin coating embedded with hard, tiny crystals called **otoliths.** When the head is tilted, gravity pulls on the heavy otoliths and the hairs bend. This induces a receptor potential. Sensory information regarding the effects of gravity (static equilibrium) is transmitted to the brain through the **vestibular nerve,** another branch of cranial nerve VIII.

o **Semicircular canals**—These are three round passages, each on a different plane. The semicircular canals have bubbles at their bases

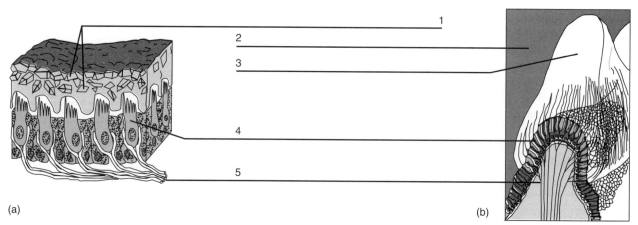

1
2
3
4
5

(a)   (b)

Figure 22.1   Identify (*a*) the structure of the macula and (*b*) the structure of the ampulla.

called **ampullae** (fig. 22.1*b*). Within each ampulla is a crest of tissue called the **crista ampullaris.** Each crista is a patch of sensory hair cells covered with a gelatinous mass (without otoliths) called the **cupula.** When the speed or direction of movement of the head changes, the inertia of the endolymph within the semicircular canals causes it to circulate. As the endolymph

circulates, it pushes the cupula and generates receptor potentials in the crista's sensory cells. The vestibular nerve carries the signal to the brain, where it is interpreted as dynamic equilibrium.

☐ **4**　Study the process of hearing illustrated in figure 22.2.
☐ **5**　Refer to what you've learned about the ear by filling in figure 22.3.

1. Sound waves strike the tympanic membrane and cause it to vibrate.

2. Vibration of the tympanic membrane causes the three bones of the middle ear to vibrate.

3. The foot plate of the stapes vibrates in the oval window.

4. Vibration of the foot plate causes the perilymph in the scala vestibuli to vibrate.

5. Vibration of the perilymph causes displacement of the basilar membrane. Short (high-pitch) sound waves cause displacement of the

basilar membrane near the oval window, and longer waves (low-pitch) cause displacement of the basilar membrane some distance from the oval window. Movement of the basilar membrane is detected in the hair cells of the spiral organ, which are attached to the basilar membrane.

6. Vibrations of the perilymph in the scala vestibuli and of the endolymph in the cochlear duct are transferred to the perilymph of the scala tympani.

7. Vibrations in the perilymph of the scala tympani are transferred to the round window, where they are dampened.

Figure 22.2　Effect of sound waves on cochlear structures.

The Ear

EXTERNAL EAR
AURICLE 1
EXTERNAL AUDITORY CANAL 2
TYMPANIC MEMBRANE 3

MIDDLE EAR
MALLEUS 4
INCUS 5
STAPES 6
AUDITORY TUBE 7

INNER EAR
COCHLEA 8
SPIRAL LAMINA 9
VESTIBULAR MEMBRANE 10
BASILAR MEMBRANE 11
COCHLEAR DUCT 12
SCALA VESTIBULI 13
SCALA TYMPANI 14
TECTORIAL MEMBRANE 15
SPIRAL ORGAN (OF CORTI) 16
COCHLEAR NERVE 17
VESTIBULE 18
VESTIBULAR NERVE 19
SEMICIRCULAR CANALS 20

Figure 22.3 Use colored pens or pencils to color the labels and the structures of the ear.

B. Hearing and Equilibrium Tests

Conduction impairment is a blocking of sound waves as they are conducted through the external and middle ear to the sensory areas of the inner ear (the *conduction pathway*). **Nerve impairment** or *sensorineural hearing loss* implies insensitivity to sound because of inherited or acquired nerve damage. The two hearing tests described are often used in clinical settings to screen patients with suspected hearing impairments.

☐ **1**   Conduct the **Rinne test** on a subject (fig. 22.4):
  o Strike a tuning fork on the base of your palm and place the handle against the subject's mastoid process.
  o When the subject no longer hears the tuning fork's hum, quickly move the prongs to the opening of the external auditory canal (but do not touch it).
  o Normally, a hum will be heard again. If not, a conductive impairment is suspected. Repeat on the other side of the head. You may want to try simulating conductive impairment in a normal subject by using swimmer's ear plugs.

☐ **2**   Conduct the **Weber test** on a subject (fig. 22.5):
  o Strike a tuning fork on the base of your palm and place the handle against the middle of the forehead.
  o Ask the subject on which, if any, side the sound seems louder.

Figure 22.4   Using a tuning fork for the Rinne test.

Figure 22.5   Weber test.

○ A conduction problem is indicated on a side that is noticeably louder than the other. Again, you may want to simulate a conduction impairment by having the subject wear one ear plug.

☐ **3** The **Romberg test** is a cerebellar function test that demonstrates a subject's ability to use information sensed by the utricle and saccule alone, without any help from the sense of vision.

○ Position the subject standing near a chalkboard or other vertical writing surface, facing away from the board.

○ Use a desk lamp or another light source to cast a shadow of the subject onto the chalkboard.

○ Have the subject stand, feet together and eyes open, for 2 minutes.

SAFETY FIRST!

If an equilibrium impairment exists, the Romberg test may result in the subject toppling over! One or two people should stand ready to catch the subject if necessary.

○ Mark the farthest edges of the subject's shadow as it sways.

○ After 2 minutes, measure the longest distance of sway.

○ Repeat with the subject's eyes closed.

○ Repeat with the subject facing perpendicular to the board.

○ Results showing excessive sway, almost to the point of falling, indicate an equilibrium problem.

LAB REPORT 22

The Ear, Hearing, and Equilibrium

Name _____

Date _____ Section _____

Specimen: *cochlea c.s.*
Total magnification: _____

Test	Left		Right	
	Normal	**Impaired**	**Normal**	**Impaired**
Rinne	☐	☐	☐	☐
Weber	☐	☐	☐	☐
Romberg	☐ Normal ☐ Impaired			

Remarks:

Figure 22.1 (*from p. 162*)

_____ 1
_____ 2
_____ 3
_____ 4
_____ 5

Put in order

_____ 1
_____ 2
_____ 3
_____ 4
_____ 5
_____ 6
_____ 7
_____ 8
_____ 9

Put in order

Arrange these structures in the order in which sound waves pass through them.

auricle

endolymph

external auditory canal

incus

malleus

perilymph

stapes

tectorial membrane/organ of Corti

tympanum

Ear divisions

———————————— 1

———————————— 2

———————————— 3

———————————— 4

———————————— 5

———————————— 6

———————————— 7

———————————— 8

———————————— 9

———————————— 10

———————————— 11

———————————— 12

———————————— 13

———————————— 14

Fill-in

———————————— 1

———————————— 2

———————————— 3

———————————— 4

———————————— 5

———————————— 6

———————————— 7

———————————— 8

———————————— 9

———————————— 10

———————————— 11

———————————— 12

———————————— 13

———————————— 14

———————————— 15

———————————— 16

———————————— 17

———————————— 18

Ear divisions

For each structure, identify whether it is part of the external ear, middle ear, or inner ear.

1. Auricle
2. Bony labyrinth
3. Cochlea
4. External auditory canal
5. Incus
6. Malleus
7. Membranous labyrinth
8. Organ of Corti
9. Saccule
10. Semicircular canal
11. Stapes
12. Tympanic membrane
13. Utricle
14. Vestibule

Fill-in

1. The ___?___ are three passages, each forming a circle and each in a different plane.
2. When traveling in an automobile, you sense that you have suddenly turned a corner. The sensation you have is mainly an aspect of the sense of ___?___ equilibrium.
3. In a weightless environment, as in deep space, the sense of ___?___ equilibrium would not work well, if at all.
4. The vestibular membrane and basilar membrane are walls of the ___?___ labyrinth.
5. The sensory patch in the utricle's lining is called the ___?___.
6. The sensory patch in the utricle's lining has receptors for the sense of ___?___.
7. The ___?___ is a passage that allows air pressure in the middle ear to reach equilibrium with atmospheric air pressure.
8. The auditory ossicles are joined to one another with ___?___ (type) joints.
9. The ___?___ has a flat, pluglike portion that fits into the oval window.
10. Both the oval window and the ___?___ window are on the boundary of the middle and inner ear.
11. During sound reception, the tectorial membrane vibrates and stimulates the hair cells in the ___?___.
12. The ___?___ is the most external portion of the external ear.
13. The ___?___ amplify sound waves by as much as 20 times their original intensity.
14. The semicircular canals have sensory structures called ___?___ within the ampullae.
15. The semicircular canals have sensory structures for the sense of ___?___.
16. The maculae of the vestibular sensory regions are covered with gelatinous material. ___?___ are embedded in this material.
17. Abnormal results in either the Weber test or Rinne test suggest a possible ___?___ impairment.
18. The Romberg test screens for problems with the sense of ___?___.

Endocrine Glands

The **endocrine system** includes a number of glands that secrete regulatory chemicals called **hormones** into the blood for distribution throughout the body. The hormones interact only with their **target cells,** cells with receptors for those hormones. Endocrinology, the study of endocrine glands and hormones, is advancing rapidly as research continues. Biologists now realize that practically every tissue in the body produces at least one hormone. However, only a handful of glands are large and specialized enough and secrete enough hormone to be considered here as part of the endocrine system. The gross and microscopic structure of some of the major endocrine glands are presented in this exercise. Exercise 24 presents topics related to endocrine hormones.

Before You Begin

☐ Read the appropriate chapter in your textbook.

☐ Set your learning goals. When you finish this exercise, you should be able to:
  o describe the principal endocrine glands and locate each in a model and in figures
  o identify histological features of some of the important endocrine glands

☐ Prepare your materials:
  o dissectible model of the human torso and head
  o microscope
  o prepared microslides:
      pituitary gland
      thyroid/parathyroid glands
      adrenal gland
      pancreas
      thymus
      pineal

☐ **Carefully** read the directions and safety tips for this exercise before starting any procedure.

A. Gross Anatomy

Using a dissectible model of the human body, locate each endocrine organ described (fig. 23.1). As you identify each gland, investigate its gross structure.

☐ **1 Pituitary**—This tiny, pea-shaped gland is located on the inferior aspect of the brain. It is cradled in the sella turcica of the sphenoid bone. The pituitary is actually two glands fused together: the **anterior pituitary** and the **posterior pituitary.**

☐ **2 Thyroid**—This single gland is located on the anterior aspect of the trachea (windpipe), near the larynx (voice box). It resembles a bow tie in that it has two lateral **lobes** and a narrow medial **isthmus** that joins them in the middle.

☐ **3 Parathyroids**—The parathyroids are tiny masses of tissue embedded in the posterior surface of the thyroid. There are usually two superior and two inferior parathyroid glands. (Many models do not represent these glands.)

☐ **4 Adrenals**—The adrenals are a pair of glands, each on the superior surface of a kidney. Each adrenal resembles a cone-shaped hat on top of a kidney.

☐ **5 Pancreas**—Both an endocrine and exocrine gland, the pancreas is a long, narrow mass of glandular tissue. Resembling a fish, the pancreas is cradled in the bend of the C-shaped duodenum (first part of the small intestine, just inferior to the stomach).

☐ **6 Thymus**—This two-lobed gland is located in the anterior mediastinum. The thymus grows until puberty, then degenerates through adulthood. It is not shown in models of the adult human body.

☐ **7 Pineal**—The tiny pineal body has already been identified as a nervous system structure in the diencephalon of the brain. Because it secretes several hormones, it is also considered an endocrine gland.

☐ **8 Testes**—The testes, the pair of male *gonads* (primary sex organs), are ovoid organs in the saclike *scrotum* external to the lower anterior trunk.

☐ **9 Ovaries**—The pair of female gonads called the *ovaries* are ovoid organs within the pelvic cavity. Each is located at the distal end of a tube leading to the uterus (womb) called a **fallopian tube.**

☐ **10** Many other organs secrete endocrine hormones, so some biologists consider them to be components of the endocrine system. For example, the digestive tract, the placenta, the kidneys, and the skin all produce endocrine hormones.

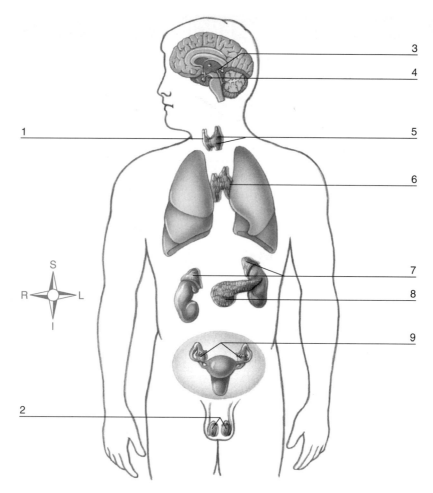

Figure 23.1 Identify the endocrine glands shown.

B. Endocrine Histology

In this activity, you are asked to identify some of the histological characteristics of endocrine glands. For each, locate the structures described in a prepared specimen and sketch your observations in Lab Report 23.

SAFETY FIRST!

Don't forget to be cautious when using the microscope and when handling prepared slides.

Landmark Characteristics

The anterior pituitary is usually more darkly stained than the posterior pituitary. Remember that you should be able to find both in a single cross section of the pituitary. The anterior pituitary has two regions: the *pars distalis* and the *pars intermedia*. Of the two, the pars intermedia is more lightly stained. The posterior pituitary has many dark, nonsecreting *pituicytes* and only a few secretory cells (which are normally not distinguishable).

☐ **1** **Pituitary**—As already stated, the pituitary is actually two glands in one. The anterior pituitary is modified epithelium, which is typical for endocrine glands. The posterior pituitary is composed of **neurosecretory tissue,** nerve tissue whose neurons secrete hormones rather than neurotransmitters. Refer to Histology Minireference Plate 12*a–b.*

☐ **2** **Thyroid**—In thyroid tissue, the glandular epithelium forms **thyroid follicles,** which contain **thyroid colloid** (the stored form of thyroid hormones). The **follicular cells,** which form the walls of each follicle, secrete the colloid. The cells between the follicles, called **parafollicular cells,** also secrete a hormone. See Histology Minireference Plate 12*c* and figure 23.2*a.*

Follicle (containing thyroglobulin) Follicular cells Parafollicular cells

Connective tissue capsule Thyroid gland

(a)

(b)

Cords of parathyroid cells (chief cells) Parathyroid gland

Figure 23.2 (*a*) Thyroid tissue. (*b*) Parathyroid and thyroid tissue.

Landmark Characteristics

One of the easiest specimens to identify, thyroid tissue has circles of cuboidal cells with uniformly stained centers. These are cross-sectional views of follicles filled with colloid. Some parafollicular cells should be visible between the circles.

☐ **3** **Parathyroids**—Parathyroid tissue can often be found in the thyroid tissue specimen because of the close association of these two gland types (fig. 23.2*b*). Parathyroid tissue does not have any follicles but a dense packing of glandular cells. **Chief cells** form cords, or rows, and are thought to be primary producers of hormone. **Oxyphil cells** are scattered among the chief cells and are considered by some to be backup hormone producers.

Landmark Characteristics

Chief cells are smaller, more densely packed cells than oxyphil cells. Oxyphil cells have more cytoplasm per cell, giving them a lighter overall appearance. As is typical in prepared histology specimens, the nuclei are darkly stained, and the cytoplasm is lightly stained.

☐ **4** **Adrenals**—Each adrenal gland is composed of an inner **medulla** and an outer **cortex.** The neurosecretory **chromaffin cells** form the medullary tissue. Cortical tissue is composed of three distinct regions of glandular epithelium:
  o **Zona glomerulosa**—The outermost region, it is just deep to the *capsule* of the organ. Here, cells are arranged in balls and loops.
  o **Zona fasciculata**—It is the middle, and thickest, region of the cortex. The cells in this zone form rather straight rows that are perpendicular to the gland's surface.
  o **Zona reticularis**—This is the innermost zone. Cells in this region form branched rows. See Histology Minireference Plate 12*d* and figure 23.3*a*.

Landmark Characteristics

With low-power magnification, you may be able to see at least part of the medulla and all the cortical zones. The medullary chromaffin cells stain more darkly than cells in the adrenal cortex. Switching to high-power magnification, you may be able to distinguish among the arrangements of cells in the three zones of the cortex. The cortex appears lighter in color than the medulla because the cortical cells do not accept the stain well.

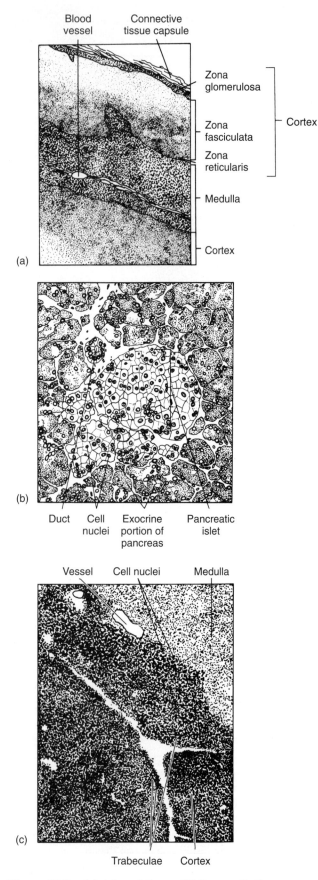

Figure 23.3   (*a*) Adrenal tissue. (*b*) Pancreatic tissue. (*c*) Thymic tissue.

☐ **5**   **Pancreas**—The pancreas can be thought of as an exocrine gland with bits of endocrine tissue scattered throughout (fig. 23.3*b*). The exocrine **acinar cells** secrete digestive *pancreatic juice.* The endocrine **pancreatic islets (of Langerhans)** are spherical masses of cells that secrete pancreatic hormones.

Landmark Characteristics

Pancreatic islets are usually easy to locate because their endocrine cells are more lightly stained than the surrounding acinar cells. Often, several islets are visible in a single low-power field.

☐ **6**   **Thymus**—The thymus has a dual role in that it functions as both a lymphatic organ and an endocrine gland. It serves as a site for the development of T-lymphocytes (immune cells) and secretes a hormone that influences lymphocyte development. The gland is surrounded by a *capsule* that extends inward, forming *trabeculae* that divide the thymus into **lobules.** Each lobule has a **cortex** largely filled with lymphocytes and a **medulla** with few lymphocytes. In the medulla, small spheres called **thymic (Hassall's) corpuscles,** whose function is unclear, can be found. See Histology Minireference Plate 14*d* and figure 23.3*c*.

Landmark Characteristics

The capsule and its inward extensions (trabeculae) are normally pale, easy-to-find structures. Near these walls, the cortex can be seen as a region of darkly stained lymphocytes. Farther away from the walls, the more lightly stained medullary region can be seen. Thymic corpuscles (circles with a nonuniform interior) can sometimes be seen.

☐ **7**   **Pineal gland**—The pineal gland is composed of neuroglia and cells sometimes called *pinealocytes,* which are thought to secrete a number of different hormones. Around puberty, calcifications called **pineal (brain) sand** may begin developing. Although several hypotheses have been proposed, the significance of brain sand is still unknown.

Landmark Characteristics

Pineal tissue appears as a hodgepodge of different cells but may have several distinct dark spots that represent calcifications ("sand").

LAB REPORT 23

Endocrine Glands

Name _____

Date _____ Section _____

Specimen: *pituitary gland*
    Total magnification: _____

Pituitary gland

Structure:

Location:

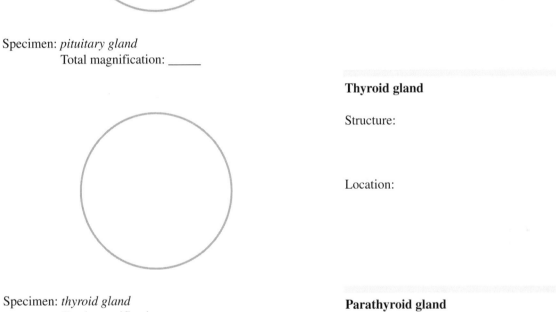

Specimen: *thyroid gland*
    Total magnification: _____

Thyroid gland

Structure:

Location:

Parathyroid gland

Structure:

Location:

Specimen: *parathyroid gland*
    Total magnification: _____

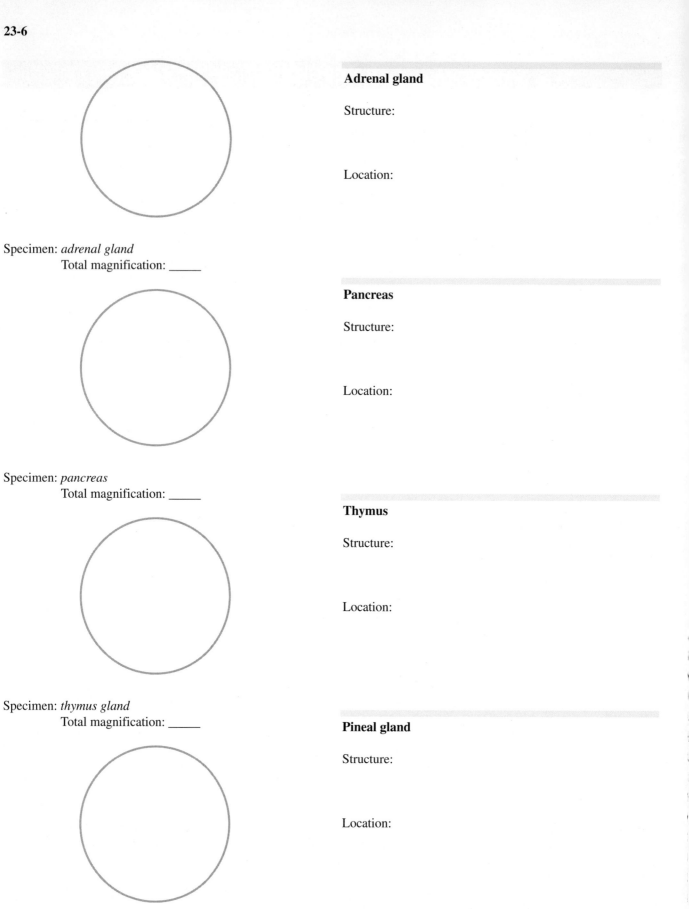

Specimen: *adrenal gland*
     Total magnification: _____

Specimen: *pancreas*
     Total magnification: _____

Specimen: *thymus gland*
     Total magnification: _____

Specimen: *pineal gland*
     Total magnification: _____

Adrenal gland

Structure:

Location:

Pancreas

Structure:

Location:

Thymus

Structure:

Location:

Pineal gland

Structure:

Location:

Hormones

Hormones are organic molecules derived from lipids or proteins. Hormones are secreted by endocrine tissues and travel throughout the body by means of the bloodstream. Once dispersed, the hormone molecules have a regulatory effect on cells that have **receptors** for those hormones. The affected cells are **target cells,** affected tissues are target tissues, and so on.

Generally, secretion of hormones is regulated by **feedback mechanisms.** *Negative feedback* mechanisms slow or stop secretion when the metabolic changes triggered in the target cells become evident or when blood hormone concentration gets too high. Negative feedback is the most commonly observed type of regulation. For example, **insulin** tends to decrease blood glucose concentration. When blood glucose is high, insulin secretion increases. When blood glucose drops, insulin secretion decreases. *Positive feedback,* less commonly observed, results in an increased hormone secretion when the metabolic changes become evident or blood hormone levels increase. For example, high levels of labor-stimulating **oxytocin (OT)** trigger uterine contractions, which push the baby and stretch the birth canal. The stretch sensation feeds back to cause more oxytocin secretion and so on. Another mechanism of regulating hormone secretion is the **biological clock.** Endocrinologists have long observed cyclic rises and falls in the secretion of several hormones related to reproduction, but more recent evidence suggests that most, if not all, hormones may be regulated to some degree by a biological time-keeping mechanism.

This activity challenges you to learn the names, sources, targets, and actions of some of the principal endocrine hormones and presents some demonstrations of hormone detection tests.

Before You Begin

☐ Read the appropriate chapter in your textbook.

☐ Set your learning goals. When you finish this exercise, you should be able to:
  o name the major hormones of the endocrine system and identify their sources, their targets, and their principal actions

  o appreciate the noncontinuous pattern of secretion in hormones
  o explain the hormonal basis of pregnancy testing
  o explain how hormone testing can predict ovulation in fertile women

☐ Prepare your materials:
  o textbook or (hormone) reference book
  o pregnancy test kit (Fact plus *pro* kit used in this example) packaged—two required per experiment
  o ovulation test kit (ClearBlue Easy kit used in this example)
  o urine samples:
      pregnant (first trimester) female
      nonpregnant female
      ovulating female (see instructions)
  o timer, watch, or clock
  o paper wipes
  o urine collection cups (disposable)
  o PHYSIOLOGY INTERACTIVE LAB SIMULATIONS (PH.I.L.S.) 4.0 and computer system (optional)

☐ **Carefully** read the directions and safety tips for this exercise before starting any procedure.

A. Endocrine Hormones

Fill in the hormone table in Lab Report 24. If you have already studied the endocrine hormones, try to fill it in without referring to a book or your notes. (If this is your first exposure to endocrine hormones, you will need to use your textbook or a reference book.) As you complete the table, you will find it a helpful study technique to shade in the rows of hormones that come from the same source with the same colored pen or pencil. A brief review of Lab Exercise 23 will be helpful.

SAFETY FIRST!

Activities B and C call for the handling of human urine specimens. Use the proper precautions for working with body fluids to protect yourself and others from the spread of disease. Use disposable, protective gloves and protective eyewear. Dispose of your specimen as instructed by the laboratory supervisor.

B. Human Chorionic Gonadotropin

Human chorionic gonadotropin (hCG) is not found on the list of hormones in Lab Report 24 because it is not secreted from one of the major endocrine glands. Recent studies suggest that tiny amounts of hCG are found in tissues throughout the adult human body. However, it seems that the only time a physiologically significant amount is secreted is from embryonic tissue in a structure called the *chorion* during the early stages of development in the womb.

As you can see in figure 24.1*a,* **estrogen** and **progesterone** levels begin to increase at **ovulation** (release of a mature *ovum*). Both hormones promote the conditions necessary for successful development of the offspring. For example, the **endometrium** (uterine lining) thickens and becomes more vascular. If fertilization and implantation do not occur, estrogen and progesterone levels fall, resulting in loss (sloughing) of the vascularized endometrial lining *(menstrual flow).* However, if pregnancy does begin, a decrease in estrogen and progesterone would result in the loss of the endometrium and the developing offspring. To prevent this loss, the offspring's tissue secretes large amounts of hCG. hCG stimulates the ovary's secretion of progesterone and estrogen, preventing the decrease in hormones that would otherwise occur. As the **placenta** develops, it begins to secrete its own progesterone and estrogen. hCG secretion decreases during the later phases of pregnancy. Because the placenta is secreting high levels of progesterone and estrogen at this time, stimulation of ovarian secretion is no longer required (fig. 24.1*b*).

High blood levels of many molecules are reflected in the urine. This is the case with the very high hCG levels during the first trimester of pregnancy. Because high levels of hCG are detectable in the urine almost immediately after fertilization of an ovum, hCG *urinalysis* is a common form of pregnancy testing.

One type of pregnancy test seldom used anymore requires that a urine sample be processed, then injected into a female laboratory animal. If the urine stimulates ovulation or similar physiological changes, it is concluded that it contains the high levels of hCG typical of early pregnancy.

A more convenient hCG test uses *monoclonal antibodies,* which are protein molecules from cultured immune cells that react with hCG (and only with hCG or similar molecules). Anti-hCG antibodies react visibly with hCG. If a reaction occurs, hCG is present in the sample (and one concludes that the subject is pregnant). If no reaction occurs, hCG must be absent or low (and the subject may not be pregnant).

Over-the-counter (OTC) pregnancy test kits and many laboratory pregnancy screening tests are based on the antibody method. Demonstrate this test by performing antibody urinalysis with the Fact plus *pro* pregnancy test kit. Any OTC test kit can be used, but this one has the advantage of an *experimental control* component. Control tests help ensure that the results are valid and not due to a side reaction or an experimental error. If a different test is used, follow the directions accompanying that test kit.

☐ **1**   Obtain two urine samples from a biological supply house: one from a pregnant woman (in the first trimester) and one from a nonpregnant woman.

Hint . . . If prepackaged urine is not used, make sure each sample is *fresh* urine from the first morning urination. Urine can be held for up to 12 hours, if kept refrigerated (not frozen). Warm the urine to room temperature before testing. If it is not the first morning urine, make sure the subject has not recently consumed a large volume of liquid. Always observe appropriate safety precautions.

☐ **2**   Arrange the test kit components as shown in figure 24.2*a*. Notice that the *test disk* plate has two round windows above the *sample well*. The larger *result window* has a pad coated with anti-hCG antibodies. The pad in the smaller *control window* does not have anti-hCG antibodies and is the basis of the control test.

☐ **3**   Use the *urine dropper* to remove urine from the *urine collection cup.*

☐ **4**   Without touching the test disk, drop three drops (only) of the urine into the *sample well.* There may be a bluish color moving across the windows of the test disk—if it occurs, this is normal.

☐ **5**   Exactly 3 minutes after dropping your sample into the *sample well,* read your result. Results may be read earlier if a red line appears in the *control window*. If a red line has not appeared in the *control window* after 3 minutes, the test is invalid and must be repeated. Do not read results after 10 minutes have passed. Figure 24.3 shows how to interpret the results.

☐ **6**   Record your results in Lab Report 24 and dispose of the materials as directed by your laboratory supervisor.

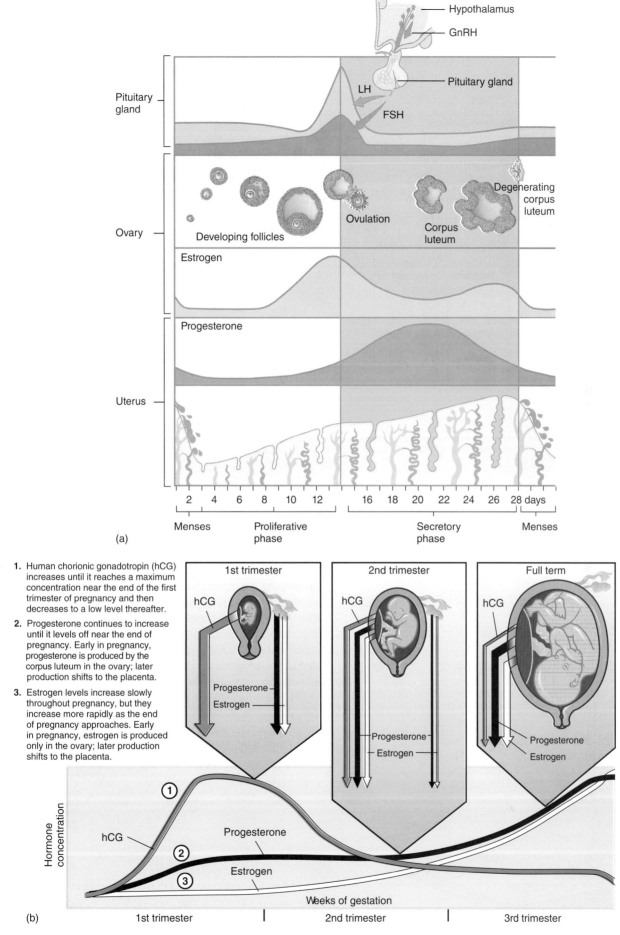

Figure 24.1 (*a*) The female reproductive cycle, assuming that the ovum is not fertilized. (*b*) Changes in reproductive hormones during pregnancy.

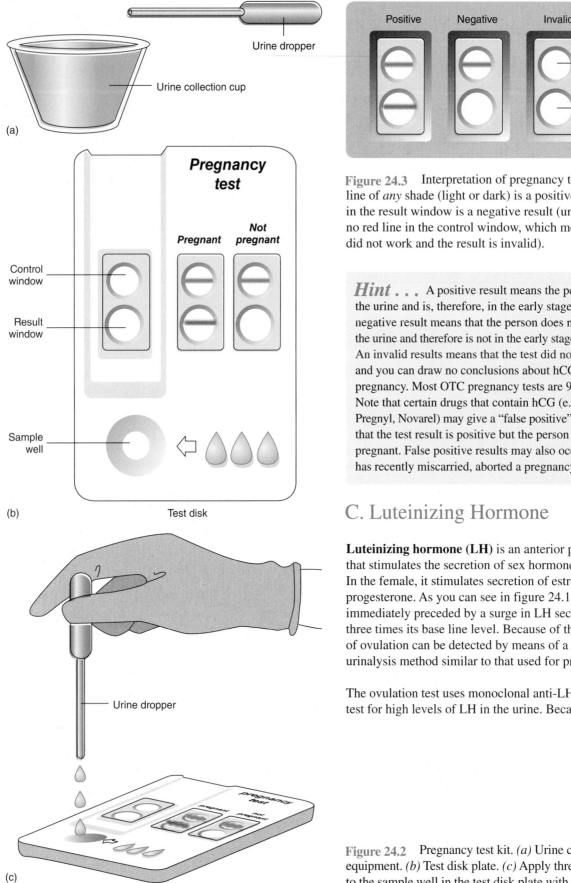

(a)

Urine dropper

Urine collection cup

Positive Negative Invalid

Control window

Result window

Figure 24.3 Interpretation of pregnancy test results. A blue line of *any* shade (light or dark) is a positive result. No line in the result window is a negative result (unless there is also no red line in the control window, which means that the test did not work and the result is invalid).

Hint . . . A positive result means the person has hCG in the urine and is, therefore, in the early stages of pregnancy. A negative result means that the person does not have hCG in the urine and therefore is not in the early stages of pregnancy. An invalid results means that the test did not work properly and you can draw no conclusions about hCG levels or pregnancy. Most OTC pregnancy tests are 99% accurate. Note that certain drugs that contain hCG (e.g., Profasi, Pregnyl, Novarel) may give a "false positive" result, meaning that the test result is positive but the person is not really pregnant. False positive results may also occur if the person has recently miscarried, aborted a pregnancy, or given birth.

C. Luteinizing Hormone

Luteinizing hormone (LH) is an anterior pituitary hormone that stimulates the secretion of sex hormones by the gonads. In the female, it stimulates secretion of estrogen and progesterone. As you can see in figure 24.1*a,* ovulation is immediately preceded by a surge in LH secretion to over three times its base line level. Because of that surge, the time of ovulation can be detected by means of a hormone urinalysis method similar to that used for pregnancy testing.

The ovulation test uses monoclonal anti-LH antibodies to test for high levels of LH in the urine. Because a series of

(b)

Pregnancy test

Pregnant *Not pregnant*

Control window

Result window

Sample well

Test disk

(c)

Urine dropper

Figure 24.2 Pregnancy test kit. (*a*) Urine collection equipment. (*b*) Test disk plate. (*c*) Apply three drops of urine to the sample well in the test disk plate with the urine dropper.

tests on the same woman must be performed over a series of 5 days to demonstrate the LH surge, it is suggested that one student from the class be selected to perform the tests at home and report the results to the class. The ideal subject already charts her ovulation cycle, has regular cycles (varying 4 days or less), and is due to ovulate soon. The chart in the ovulation test kit can help determine a good candidate, and the ideal time, for testing.

Hint . . . The instructions here apply to the Clearblue Easy ovulation test kit. Use the directions that accompany the particular test kit that you are using.

☐ **1** For each of 5 days on which the test will be performed, use the following procedure. For best results, collect urine at the *same time each day*.

☐ **2** Remove the *cap* from the Clearblue Easy test device (fig. 24.4*a*) and, with the tip pointing downward, immerse the *absorbent sampler* for 20 seconds in fresh urine collected in a clean, dry urine collection cup *or* hold the *absorbent sampler* in the urine stream for 5 seconds (only). Refer to figure 24.4*b*.

☐ **3** Still keeping the tip pointing downward, remove the absorbent sampler from the urine and replace the cap.

☐ **4** After 3 minutes, read the result of the test in the *result window* of the test device. Use figure 24.5 to read the test result.

☐ **5** Record your results over a span of 5 days in Lab Report 24 and dispose of the materials as directed by your laboratory supervisor or in the test kit.

Figure 24.4 Ovulation (LH) test kit. *(a)* Clearblue Easy ovulation test kit for measuring LH in urine. *(b)* The tip of the absorbent sampler should be held pointing downward and immersed in the urine stream or in urine collected in a urine collection cup.

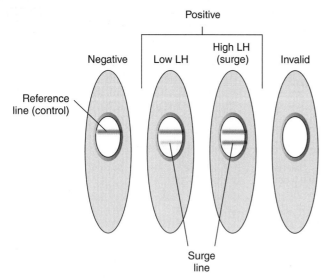

Figure 24.5 Interpretation of ovulation (LH) test results. A blue surge line of *any* shade (light or dark) is a positive result (meaning that LH is present in the sample). If the surge line is as dark as the reference line, then LH is at the peak of its surge and ovulation will follow in 24 to 36 hours. No surge line is a negative result (unless there is also no blue reference line, which means that the test did not work and the result is invalid).

Advanced Scientific Inquiry

PHYSIOLOGY INTERACTIVE LAB SIMULATION (PH.I.L.S. 4.0)

If available, use the computerized laboratory simulation PH.I.L.S. 4.0 to simulate the following activity that explores hormone function:

• *Thyroid Gland and Metabolic Function (simulation #19)*

• *Insulin and Glucose Tolerance (simulation #20)*

For each of the activities that you perform, print the lab report according to the directions given in the activity and attach a copy of all sheets to your lab report (or as your instructor directs).

Hint . . . A negative result (dark blue *reference line* and no *surge line*) means that LH was not detected in the subject's urine and therefore the person is not close to the time of ovulation. A positive result means that there is LH in the urine. If the *surge line* is present but is a lighter shade of blue than the *reference line,* then the result is positive, but not near the time of ovulation. If the *surge line* is as dark as the *reference line,* then the LH is at the peak of its surge (see fig. 24.1*a*[*top*]) and therefore ovulation should occur in the next 24 to 36 hours. The *surge line* returns to a lighter blue shade after the LH surge peak and then to no surge line at all. If the dark blue *reference line* does not appear, then the test did not work and the result is invalid.

Misleading results may occur if the subject is taking fertility drugs containing LH or hCG. The drug Clomid does not affect the test, but it may lengthen the reproductive cycle so that more than 5 tests must be performed to confirm the LH surge. This also applies to women who have irregular cycles for other reasons, such as the recent use of birth control pills.

Hormones

Name _____

Date _____ Section _____

Use your textbook or class notes to fill in this table.

HORMONE TABLE			
Hormone	Gland (Source)	Target	Action
Antidiuretic hormone			
Oxytocin (OT)			
Growth hormone (GH)			
Thyroid-stimulating hormone (TSH)			
Adrenocorticotropic hormone (ACTH)			
Melanocyte-stimulating hormone (MSH)			
Luteinizing hormone (LH)			
Follicle-stimulating hormone (FSH)			
Prolactin (PRL)			
Thyroid hormone (T_3/T_4)			
Calcitonin (CT)			
Parathyroid hormone (PTH)			
Epinephrine (Epi)			
Aldosterone			
Cortisol			

Hormone	Gland (Source)	Target	Action
Melatonin			
Thymosin			
Insulin			
Glucagon			
Testosterone			
Estrogen			
Progesterone			
Prostaglandin(s)			

DATA TABLE					
Test	Sample	Color Result	Color Control	Result	Explanation
Pregnancy (hCG)	Pregnant			hCG ☐ present (+) ☐ not present (−)	
	Nonpregnant			hCG ☐ present (+) ☐ not present (−)	
Ovulation (LH)	Day 1			LH ☐ high (+) ☐ low (+) ☐ not present (−)	
	Day 2			LH ☐ high (+) ☐ low (+) ☐ not present (−)	
	Day 3			LH ☐ high (+) ☐ low (+) ☐ not present (−)	
	Day 4			LH ☐ high (+) ☐ low (+) ☐ not present (−)	
	Day 5			LH ☐ high (+) ☐ low (+) ☐ not present (−)	

Discussion Questions

Answer these questions on a separate sheet of paper.

1. Explain how hCG secretion is regulated. Is it secreted by a pregnant woman or her offspring?

2. hCG depresses some reactions of the immune system. What adaptive advantage do you think this has?

3. How is LH secretion regulated? What effects does the changing blood levels of this hormone have?

4. Compare and contrast the reproductive roles of LH and hCG.

LAB EXERCISE 25

Blood

Blood is the fluid tissue that circulates within the cardiovascular system. Typical of connective tissues, it has a dominant extracellular matrix. In this case, the matrix is a liquid called **plasma.** Suspended in the plasma are blood cells, or **formed elements.** The characteristics of each person's blood vary with age, sex, metabolic condition, health condition, genetics, and other factors. The approximate values for the components of blood in a normal adult are shown in figure 25.1. Clinical evaluation of blood characteristics is important in assessing the condition of patients. This exercise provides demonstrations of just a few of the more common tests. Many blood tests are done by automated machines, at least in some areas, but it is important to understand the bases of these tests to interpret them correctly.

Before You Begin

☐ Read the appropriate chapter in your textbook.

☐ Set your learning goals. When you finish this exercise, you should be able to:
   o name the components of human blood tissue and identify their functions
   o demonstrate how common blood tests are performed
   o interpret the meaning of important blood tests
   o state the limitations of clinical blood testing

☐ Prepare your materials:
   o blood sample (see SAFETY FIRST! box)
   o Autolet blood drawing kit or sterile lancets
   o alcohol pads
   o paper towels, lab wipes
   o puncture-proof BIOHAZARD container
   o disinfecting solution
   o clean microscope slides
   o bibulous paper
   o distilled water (in dropper bottle)
   o Wright stain (in dropper bottle)
   o wax pencil
   o heparinized capillary tubes
   o microhematocrit centrifuge
   o capillary tube sealing clay or plugs
   o metric ruler (mm)
   o Tallquist test paper and Hb scale (or hemoglobinometer kit)
   o toothpicks
   o blood typing sera: *anti-A, anti-B, anti-D*
   o slide warming box
   o microscope
   o PHYSIOLOGY INTERACTIVE LAB SIMULATIONS (PH.I.L.S.) 4.0 and computer system (optional)

☐ **Carefully** read the directions and safety tips for this exercise before starting any procedure.

SAFETY FIRST!

This exercise calls for the use of human blood samples. These tests should be performed as a demonstration by the instructor using packaged, contaminant-free blood or a blood substitute. If students are to perform the tests with fresh blood, they should handle *only their own blood.* Any surface that comes into contact with blood *or MAY have been accidentally spattered* (including the Autolet unit) should be cleaned with a suitable disinfecting agent IMMEDIATELY. All used blood samples and disposable items, including wipes and towels, are to be discarded immediately in the appropriate BIOHAZARD container. Follow current laboratory practice by wearing a lab coat, gloves, and protective eyewear while in the lab.

A. Sampling Blood

Blood sampling for clinical testing usually requires only tiny amounts of blood, often only a few drops. The usual method for collecting small blood samples is to draw it from a punctured finger. This activity tells you how blood can be drawn for clinical testing. Follow these steps if you are collecting your own blood.

Hint . . . If you are using your own blood for tests in this exercise, it may be best to draw all the blood needed for various tests at one time. Therefore, make sure that you have all your equipment standing by so that the blood can be collected and processed immediately. You may wish to collect blood in several heparinized capillary tubes.

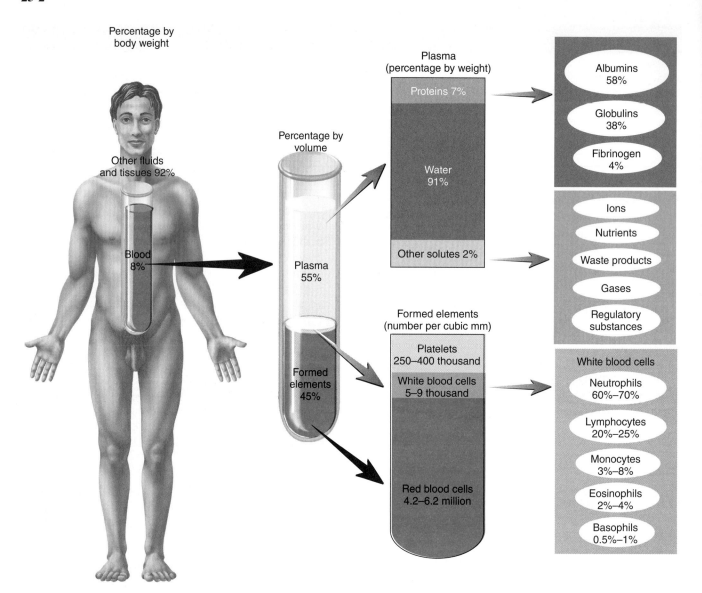

Figure 25.1 Composition of blood.

☐ **1** Clean your hands thoroughly with disinfecting hand soap (which should always be available at the lab sinks). Prepare the skin of the puncture site (see step 2) by dabbing it with a fresh alcohol pad. Do not use the pad again.

☐ **2** If you are using the Autolet kit, follow the directions in the kit and disinfect the entire unit after each use. If you are using a lancet, puncture the palmar surface of your nondominant hand's fourth finger. Use a quick, firm stroke.

☐ **3** Holding the arm down, wipe away the first drop of blood with a lab wipe. Immediately begin collecting drops of blood required for each test. *Do not squeeze the finger to draw more blood, because that will slow circulation through the finger.* If the blood stops flowing, you may have to puncture another finger. Begin at step 1 and *do not reuse the lancet or any other disposable item,* even if you are using the same finger. Dispose of all remnants of the blood collection in the labeled BIOHAZARD container.

B. Blood Smear

Investigate blood tissue by making a **blood smear** (fig. 25.2). As its name implies, it is merely some blood smeared on a slide, then prepared for microscopic examination.

(a)

(b)

(c)

(d)

Figure 25.2 Preparing a blood smear. (*a*) Drop blood near the end of the slide. (*b*) Spread the blood in a line across the slide. (*c*) *Pull* the blood across the slide to form a smear. (*d*) Mark just outside the edges of the smear with a wax pencil (to contain the stain).

☐ **1** Place a drop of blood (from a finger or capillary tube) on a clean microscope slide (about 2 cm from one end, as in figure 25.2*a*).

☐ **2** Hold the end of another clean slide at a 50° angle against the first slide and push the blood drop away from the middle of the first slide. Figure 25.2*b* shows this move and how the blood drop spreads out to form a line of blood where the two slides meet. Now push the second slide toward the middle of the first slide. This makes an even smear across the first slide. Stop before reaching the end of the slide. Put the spreader slide directly into the BIOHAZARD container.

☐ **3** When the blood has dried, draw two lines on the sample slide with a wax pencil, as shown in figure 25.2*d*. Mark on clean glass, not through the smear. Drop a few drops of Wright stain on the smear, counting them as you do so. After 4 minutes, add an equal number of drops of distilled water to the smear. After 10 minutes, gently rinse the slide under tap water for 30 seconds; then blot it dry with bibulous paper.

☐ **4** Scan the slide under low power on your microscope until you find an area where the cells are thinly spread. Switch to high power and observe the blood cells on the slide. Do all the cells look the same? Are different cells (if there are any) distributed equally? What unique characteristics do these cells have (compared with other cells you have observed)? Describe the pattern or order of the cells in the tissue observed. Write your answers to these questions, and any other observations, in Lab Report 25.

Hint . . . If your microscope has an oil-immersion objective, you may want to use it to see the blood cells more clearly. After focusing on high dry, rotate the nosepiece *halfway* to the high-oil objective. Put a drop of immersion oil on the slide; then rotate the high-oil objective into the oil drop. You may need to increase the lighting.

C. Differential White Blood Cell Count

The larger, nucleated cells that you may have observed in activity B are **white blood cells,** or **leukocytes.** White blood cells are far less numerous than **red blood cells (erythrocytes)** and **platelets (thrombocytes).** Figure 25.3 shows the different types of blood cells. White blood cells have numerous protective functions in the body, so their numbers sometimes change in response to changes in health. Different leukocytes have different functions, so the changes may be reflected in changes in the relative proportions of different types of leukocytes. Clinically, information about

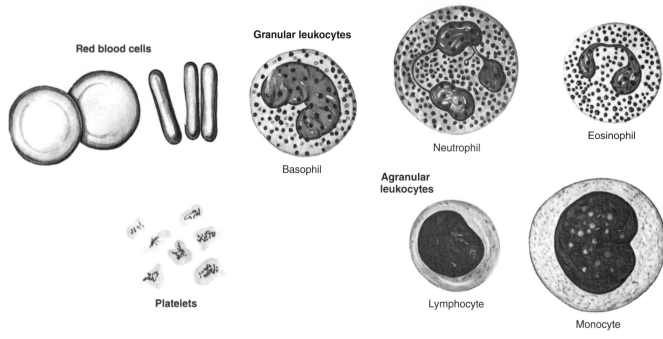

Figure 25.3 Representative human blood cells.

relative proportions of leukocyte types is very useful. A test that gives this information is the **differential white blood cell count:**

☐ **1** Focus on a portion of the blood smear, as in activity B. Look around the field for any leukocytes.

Landmark Characteristics

White blood cells (WBCs) have nuclei and are distinctly larger than red blood cells (RBCs). **Agranular leukocytes** have no granules (spots) visible in the cytoplasm, but **granular leukocytes** do.

- **Monocytes** are very large agranular leukocytes with large, variably shaped nuclei.
- **Lymphocytes** are agranular WBCs that are almost as small as RBCs. Lymphocyte nuclei are sometimes so large that they appear to have no cytoplasm.
- **Neutrophils** are granular WBCs that have small, pinkish granules and lobed nuclei that resemble several links of sausage.
- **Eosinophils** have red granules and two-lobed, dark nuclei.
- **Basophils** have fewer granules, which are a bluish tint and of variable size. Basophils have large, two-lobed or kidney-shaped nuclei.
- See Histology Minireference Plate 15 and figure 25.3.

☐ **2** In Lab Report 25, mark off how many leukocytes you see of each type until you reach 100 total cells. Obviously, you'll have to move around to different fields to reach 100 cells. Move the slide in the manner shown in figure 25.4 so that you don't count the same portion of the slide twice.

Figure 25.4 Move the slide in this manner to avoid counting the same cells more than once.

☐ **3** Determine the total number of WBCs of each type, the total for all types being 100. This number is also the percentage (because 100 is the total, 1 cell is also 1% of the cells).

Normal Blood Values

Monocytes: 3% to 8% (↑ in chronic infections)

Lymphocytes: 20% to 25% (↑ in antibody reactions)

Neutrophils: 60% to 70% (↑ in acute infections)

Eosinophils: 2% to 4% (↑ in allergic reactions)

Basophils: 0.5% to 1% (↑ in chronic infections)

□ **4**   How do your results compare with the normal values given in the box?

D. Hematocrit

The blood is composed of a number of components. The plasma is the extracellular fluid matrix of blood tissue. Mostly water, it also contains ions (e.g., Na^+, Cl), **plasma proteins** (e.g., *antibodies, albumin*), and other dissolved substances (e.g., *urea*, O_2, *glucose*). The formed elements (blood cells, or *blood solids*) include RBCs, WBCs, and platelets. You've already completed one blood test that looks at WBCs. The blood test demonstrated in this activity explores the relative proportions of blood solids and blood plasma.

The **hematocrit** test is also called the **PCV** (packed cell volume) determination because it involves the packing of all the cells in a blood sample at one end of a tube. The sample is placed in a *centrifuge,* a machine that spins a tube of blood and allows centrifugal force to push the heavy blood cells to the bottom. Once the cells are packed, the proportion of cells to total volume can be calculated. The results are used to determine whether there is a shortage of RBCs (as in some *anemias*), an excess of RBCs (*polycythemia*), or dehydration.

□ **1**   Fill a heparinized capillary tube about 75% full of blood by touching it to the blood drop on your punctured finger (fig. 25.5*a*). Heparinized tubes are red-tipped for easy identification.

□ **2**   Seal the unused end of the tube by holding it near the tip and pushing it into a tray of sealing clay at a 90° angle with a twisting motion (fig. 25.5*b*). A plug of clay will then seal the tube. Remove the tube. Plastic plugs for this purpose are also available.

SAFETY FIRST!

Capillary tubes are easily broken, causing injury and contamination by the blood. Have the instructor demonstrate the proper technique for sealing them before attempting it yourself. Handle capillary tubes with gloved hands.

□ **3**   Place the tube in one of the numbered grooves of a microhematocrit centrifuge with the sealed end facing the perimeter of the centrifuge tray (fig. 25.5*c*). Close the top of the centrifuge according to the manufacturer's instructions and spin the tubes for 4 minutes.

□ **4**   Remove your tube from the centrifuge and notice that the RBCs are now packed in one end of the tube. There may be a thin, lightly colored *buffy coat* topping the RBCs. This coat is actually the WBCs, which are slightly less dense than the RBCs. The rest of the tube contains the yellowish, transparent plasma.

Figure 25.5   Hematocrit determination. (*a*) Draw blood into a red-tipped capillary tube. (*b*) Seal the unused end by twisting it in a tray of sealing clay. (*c*) Place the tube, sealed end outward, in a centrifuge. (*d*) Measure the total length and the packed cell length—in this example, (41 ÷ 79) × 100 = 52%.

☐ **5** Lay the tube on a paper towel next to a metric ruler and measure the total length of the sample, from the clay-RBC border to the end of the plasma. Next, measure only the RBC portion. Report your results (to the nearest mm) in Lab Report 25.

☐ **6** Divide the length of the RBC section by the total sample length. Multiply your answer by 100 to arrive at the percent volume of packed red blood cells. This is the hematocrit value, reported with or without the % symbol. Use the normal values given in the box to determine whether your results fall within the normal range.

Normal Blood Values

Hematocrit: male, 40% to 54%; female, 38% to 47%

↑ in severe dehydration, shock, polycythemia

↓ in anemia, leukemia, cirrhosis, hyperthyroidism

E. Hemoglobin Determination

Hemoglobin is the pigment inside RBCs that has an affinity for O_2 (and CO_2), so it transports blood gases. Although knowing the number or proportion of RBCs in the blood may hint at the total hemoglobin (Hb) content, a *hemoglobin determination* gives a relatively accurate figure. Most methods operate on the principle that, the denser the color of the sample, the higher the content of Hb pigment. Of course, the Hb first has to be liberated from the RBCs and allowed to diffuse evenly throughout the plasma. The more accurate methods use *colorimeters* or *spectrophotometers* that measure the percentage of filtered light transmitted through the sample. Some clinics use a handheld version of this type of instrument called a *hemoglobinometer*. A less accurate method, the *Tallquist method,* is used here to demonstrate hemoglobin determination. (If you have a hemoglobinometer and wish to use it, follow the directions supplied in the accompanying kit.)

☐ **1** Place a drop of blood on the special paper in the Tallquist test kit. The RBCs break open in the paper.

☐ **2** Allow the drop to dry long enough to lose its glossy appearance but not long enough for it to become brown.

☐ **3** Compare the color of the blood spot with the colors in the Tallquist chart in your kit. Determine the Hb content by matching colors as closely as possible. Record your result in grams of Hb per 100 mL of blood (g/100 mL) in Lab Report 25. How does it compare with normal values?

Normal Blood Values

Hb: male, 14 to 16.5 g/100 mL; female, 12 to 15 g/100 mL

↑ in polycythemia, congestive heart failure, obstructive pulmonary disease, high altitudes

↓ in anemia, hyperthyroidism, liver cirrhosis, severe hemorrhage

---

F. Blood Typing

All cells have different proteins, on the surface of their cell membranes, that act as identification tags. The human immune system has cells and chemicals that can recognize proteins as *non-self proteins.* In this way, immune processes can try to destroy or inhibit proteins (and cells to which they may be attached) that are foreign to a person's body. The identifying proteins on cell surfaces are determined by heredity. If a foreign tissue is transplanted into someone's body, as in blood transfusions, the immune system will destroy new blood cells that don't have the *self protein* marker.

Advanced Scientific Inquiry

If available, use the computerized laboratory simulation PH.I.L.S. 4.0 to simulate how blood types are analyzed. This can be done as pre-lab to Activity F or as alternative to performing Activity F.

• *Blood Typing (simulation #36)*

After performing the simulation activity, print the lab report according to the directions given in the activity and attach a copy of all sheets to your lab report (or as your instructor directs).

To prevent this *tissue rejection* from happening, biologists have determined which blood cell marker proteins elicit life-threatening immune reactions. Biologists have also devised tests to determine the presence of these proteins and systems of naming them. This demonstration shows you how some of these **blood typing** tests work.

Before beginning the demonstration, you must learn some new terms. The term **antigen** refers to a molecule that elicits an immune response. In the case of blood transfusions, marker proteins on the donor's RBCs can be antigens (if they are non-self relative to the recipient). The term **antibody** refers to a type of plasma protein produced by the immune system. Antibodies react with non-self markers (antigens) and try to destroy or inhibit them. One type of antigen-antibody reaction is the **agglutination reaction,** in which antigens are "glued" together with antibodies, like flies on flypaper. This clumping (agglutination) is the reaction commonly seen when incompatible blood types are transfused.

In the **ABO system** of blood typing (fig. 25.6), two blood antigens are important: the **A antigen** and the **B antigen.** People with *type A* blood have the A antigen but not the B antigen. *Type B* blood has the B antigen but not the A antigen. *Type AB* blood has both A and B antigens, whereas *type O* blood has neither A nor B antigens. To type blood in the ABO system, blood is mixed with *anti-A serum* and *anti-B serum.* Serum is plasma with the clotting factors removed. Anti-A serum contains antibodies that cause agglutination when A antigen is present. Anti-B serum contains antibodies that react with the B antigen.

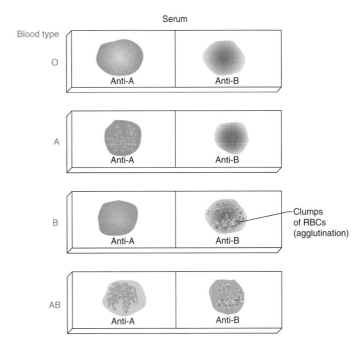

Figure 25.6 Possible results in the ABO typing test.

SAFETY FIRST!

Handle blood serum with the same caution that you would use when handling any blood sample.

☐ **1** Place a clean microscope slide on a paper towel and write "Anti-A" near the left side and "Anti-B" near the right side. Use a wax pencil to make two 1 cm circles on the slide, one on the right and one on the left.

☐ **2** Put a drop of fresh blood in each circle.

☐ **3** To the drop of blood in the left circle, add a drop of anti-A serum *without touching the dropper to the blood.* Similarly, add a drop of anti-B serum to the right blood drop. Using a different toothpick for each circle, mix the serum and the blood.

□ **4**   Check each mixture for a clumping (agglutination) reaction: the mixture changes from a uniform reddish color to distinct, dark clumps in a transparent medium. If a reaction occurred in the left circle (anti-A), then the A antigen is present. If no reaction occurred, then the antigen is not present. According to the ABO system, what type is the blood (A, B, AB, or O)? See figure 25.6 for possible results.

Another commonly used blood typing system is the **Rh system,** which deals with the **Rh antigen.** Sometimes called the *D antigen,* this blood cell marker does not elicit an immune reaction in someone who doesn't have it until the second exposure to the antigen because the immune system doesn't make **anti-D antibodies** until it has been exposed to the

D antigen. Once exposed, however, an army of antibodies (and anti-D–producing cells) stands ready to react with the D antigen. A person with the antigen is said to be *Rh positive* and a person without this antigen is termed *Rh negative.* The Rh typing test is similar to the ABO test:

□ **5**   Draw a circle on a clean glass slide and place a drop of blood within it (as described in steps 1 and 2).

□ **6**   Add a drop of Anti-D serum and place the slide on a slide warmer (fig. 25.7). This device warms the slide to about 45°C and tilts it back and forth. Because the Rh reaction is less severe than reactions in the ABO system, such treatment is necessary to help the reaction so that it can be easily observed. Look for fine clumping. What Rh type is the sample?

Mixture of blood and anti-D serum

Plate heated to 45°C (113°F)

Rh TYPING BOX

Figure 25.7   Slide warming box for Rh blood typing.

LAB EXERCISE 25

Blood

Name _____

Date _____ Section _____

Specimen: *blood smear*
Total magnification: _____

Observations: *blood smear*

Differential WBC Count Tally Grid					
Type:	**Monocyte**	**Lymphocyte**	**Neutrophil**	**Eosinophil**	**Basophil**
Tally marks:					
Totals:					
Percentages:					
Normal value:					
Interpretation:					

Blood Test Results

Subject:			Sex: ☐ male ☐ female
Test	**Normal value**	**Result**	**Interpretation**
Hematocrit			
Hemoglobin METHOD: _____			
ABO type			
Rh type			

Fill-in

_____ 1
_____ 2
_____ 3
_____ 4
_____ 5
_____ 6
_____ 7
_____ 8
_____ 9
_____ 10
_____ 11
_____ 12
_____ 13
_____ 14
_____ 15
_____ 16
_____ 17
_____ 18
_____ 19
_____ 20

Fill-in

1. A blood cell protein that elicits an immune reaction may be called a(n) __?__.
2. The __?__ scale compares the color of blood-soaked paper with a standard, allowing one to determine hemoglobin content.
3. Lab materials soiled by a blood sample should be disinfected or disposed of in a puncture-proof __?__ container.
4. __?__ stain is used to make the cells in a blood smear more easily observable.
5. Blood is often collected in __?__ capillary tubes, rather than plain capillary tubes, to prevent clotting of the blood.
6. A(n) __?__ is a machine required for the hematocrit test.
7. The PCV test determines the ratio of __?__ volume to total blood volume.
8. __?__ is an oxygen-carrying pigment in the blood.
9. A person with a high differential count of __?__ WBCs is suspected of having an acute infection.
10. A person with a high differential count of __?__ WBCs is suspected of having a chronic infection.
11. Robert's hematocrit is 53%. This is normal/abnormal?
12. Maria's hematocrit also is 53%. This is normal/abnormal?
13. On a very hot day, Hiro has been outside doing heavy work and just collapsed. If you took a blood sample right now, his hematocrit would be higher/lower? than his usual hematocrit value.
14. The hemoglobinometer is more/less? accurate than the Tallquist method.
15. Fred has just been informed that his Rh type is negative. This is normal/abnormal?
16. Fred has type AB- blood. He needs a transfusion, and his wife (type AB+) is willing to donate her blood. Assuming Fred has never had any type of blood transfusion before, it is likely/unlikely? that there will be an antigen-antibody reaction if the transfusion is done.
17. Kevin has B+ blood. His plasma is likely to contain anti-__?__ antibodies.
18. Bruce has been living in the mountains for years. The hemoglobin content of his blood is likely to have risen/fallen? since he moved there.
19. Irma has cirrhosis of the liver. Her Hb value is likely to be high/low?
20. Hb values are usually reported in grams of Hb per __?__ of blood.

Structure of the Heart

The **heart** is a four-chambered, hollow organ primarily composed of cardiac muscle tissue. It contracts rhythmically, pumping blood into the **arteries.** After passing through tissues, blood returns to the heart by way of the **veins** and is pumped again. This exercise challenges you to explore the anatomy of the heart through the use of models and preserved specimens.

Before You Begin

☐ Read the appropriate chapter in your textbook.

☐ Set your learning goals. When you finish this exercise, you should be able to:
  ○ describe the structure of the heart
  ○ locate anatomical features of the heart in models and in preserved mammalian specimens
  ○ explain the function of major heart structures

☐ Prepare your materials:
  ○ dissectible models of the human heart
  ○ preserved sheep heart
  ○ dissection tools and trays
  ○ wooden dowels (1 cm diameter × 12 cm), pencils, or dull probes

☐ **Carefully** read the directions and safety tips for this exercise before starting any procedure.

A. Human Heart Anatomy

Using dissectible models and the aid given in this exercise, find these features of the heart:

☐ **1** Identify these structures on the external aspect, ventral surface:
  ○ **Interventricular sulcus**—This diagonal groove is located between the walls of the two lower heart chambers (**ventricles**). Along this groove lie the **anterior interventricular artery** and the **great cardiac vein.**
  ○ **Auricles**—These are the flaplike outpouchings of the left and right **atria** (the upper heart chambers).
  ○ **Atrioventricular sulci**—These are grooves between the walls of the atria above and the ventricles below. Locate the **small cardiac vein**

and **right coronary artery** on the right and the **great cardiac vein** and **circumflex artery** on the left.
  ○ **Aorta**—The largest artery of the body, it forms the **aortic arch** above the heart.
  ○ **Pulmonary artery**—Somewhat smaller than the aorta, this vessel leaves the heart as a single **trunk** but soon branches to become the **left** and **right pulmonary arteries.**
  ○ **Superior** and **inferior vena cava**—These two large veins communicate with the right atrium.
  ○ **Apex**—The apex is the lower "point" of the heart.

☐ **2** Identify these features of the heart on the external aspect, dorsal surface:
  ○ Atria—These are the upper left and right chambers. They have relatively thin walls.
  ○ Ventricles—These are the lower left and right chambers. They have relatively thick walls.
  ○ Interventricular sulcus—It is similar to that on the ventral surface. Locate the **middle cardiac vein** and the **posterior ventricular artery.**
  ○ **Pulmonary veins**—These veins communicate with the atria.

☐ **3** Identify these features visible on the internal aspect:
  ○ Atria—They are distinguished by their position and thin walls.
  ○ Ventricles—They are thick-walled lower chambers. Note that one ventricle has thicker walls than the other. What functional adaptation does this represent?
  ○ **Interventricular septum**—This heart wall separates the left and right ventricles from each other.
  ○ **Cuspid valves**—Also called **atrioventricular (AV) valves,** these valves ensure one-way flow of blood from the atria into the ventricles. The left AV valve, or **mitral (bicuspid) valve,** is composed of two cusps (flaps). The right AV valve, or **tricuspid valve,** has three cusps. Each cusp is attached to the wall of the ventricle below by means of fibrous **chordae tendineae** connected to fingerlike projections of the ventricular myocardium called **papillary muscles.**
  ○ **Semilunar (SL) valves**—The right SL valve, or **pulmonary semilunar valve,** ensures one-way flow from the right ventricle into the pulmonary

Figure 26.1 (*a, b*) Ventral view of the sheep heart. (*c, d*) Anterior view of the sheep heart, analogous to a superior view of the human heart.

artery. The left SL valve, or **aortic semilunar valve,** is at the entrance of the aorta. Semilunar valves are each composed of thin-walled bags that hang from the walls of the vessel.

o **Myocardium**—This is the muscular layer of the heart wall.

o **Endocardium**—The thin endothelial lining of the heart chambers, it covers the beamlike **trabeculae** on the inner face of the myocardium.

B. Sheep Heart Dissection

The sheep heart is very similar in structure to the human heart. It is nearly the same size, so it makes an ideal study specimen.

SAFETY FIRST!

Observe the usual safety precautions when dissecting your specimen. Heed the safety advice accompanying the preservative and avoid cuts and punctures when using the dissecting tools. As always, dispose of your specimen as instructed.

□ 1 Orient yourself to the specimen. Using figure 26.2 as a guide, locate the dorsal and ventral surfaces. (The shape of your specimen may have become distorted in shipping, so don't rely on shape as a guide.) Recall that directions for the sheep heart are based on the fact that the sheep is a four-legged animal, so its *heart is oriented differently than in the human.*

□ 2 Identify the structures of the external aspect of the sheep heart as you did for the human heart. Some adipose tissue may have to be removed so that you can see all the structures clearly. You may not be able to locate all of the *coronary vessels* because they are buried under fat. Many of the large heart vessels may have been cut very closely to the heart wall, so they appear as holes or short stubs. Use a wooden dowel or pencil to open the vessels for better viewing (as in figs. 26.1*a* and 26.2*a*). Some identifications of external structures will be tentative until you open the heart and verify your observations.

□ 3 Use a long knife or your scalpel and scissors to cut a frontal section in your specimen (as in fig. 26.2*a*). Try to identify the internal heart features as you did with the human heart model.

□ 4 Review what you've learned about the structure of a human heart by completing figure 26.3.

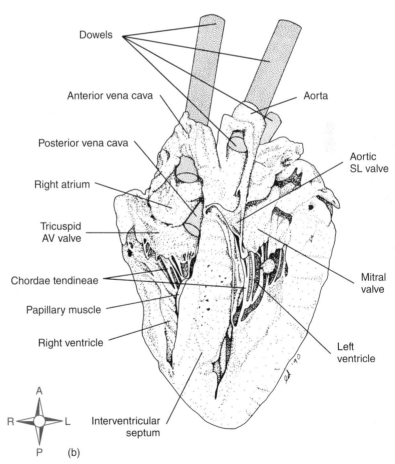

Figure 26.2   Ventral view of a frontal section of the sheep heart.

The Heart

RIGHT ATRIUM 1
LEFT ATRIUM 2
RIGHT VENTRICLE 3
LEFT VENTRICLE 4
INTERVENTRICULAR SULCUS 5
ANTERIOR INTERVENTRICULAR
   ARTERY 6
GREAT CARDIAC VEIN 7
SMALL CARDIAC VEIN 8
RIGHT CORONARY ARTERY 9
CIRCUMFLEX ARTERY 10
LEFT CORONARY ARTERY 11
AORTA 12
PULMONARY ARTERY 13
SUPERIOR VENA CAVA 14
INFERIOR VENA CAVA 15
INTERVENTRICULAR SEPTUM 16
MYOCARDIUM 17
EPICARDIUM 18
MITRAL VALVE 19
TRICUSPID VALVE 20
PULMONARY VEIN 21
CHORDAE TENDINEAE 22
PAPILLARY MUSCLE 23
AORTIC SEMILUNAR VALVE 24
PULMONARY SEMILUNAR VALVE 25

Figure 26.3 Use contrasting colored pens or pencils to color the labels and these figures of the heart.

Name _____

Structure of the Heart

Date _____   Section _____

Use this table as a checklist for your study of the heart. Don't forget to fill in the function(s) column.

Structure	Sheep	Human	Function(s)
Right atrium	☐	☐	
Right auricle	☐	☐	
Left atrium	☐	☐	
Left auricle	☐	☐	
Right ventricle	☐	☐	
Left ventricle	☐	☐	
Interventricular sulcus	☐	☐	
Anterior interventricular artery	☐	☐	
Great cardiac vein	☐	☐	
Small cardiac vein	☐	☐	
Right coronary artery	☐	☐	
Circumflex artery	☐	☐	
Left coronary artery	☐	☐	
Aorta	☐	☐	
Pulmonary artery	☐	☐	
Superior vena cava	☐	☐	
Inferior vena cava	☐	☐	
Interventricular septum	☐	☐	
Myocardium	☐	☐	
Epicardium	☐	☐	
Mitral valve	☐	☐	
Tricuspid valve	☐	☐	
Chordae tendineae	☐	☐	
Papillary muscle	☐	☐	
Aortic semilunar valve	☐	☐	
Pulmonary semilunar valve	☐	☐	

26-6

Put in order

_____ 1

_____ 2

_____ 3

_____ 4

_____ 5

_____ 6

_____ 7

_____ 8

_____ 9

_____ 10

_____ 11

_____ 12

Fill-in

_____ 1

_____ 2

_____ 3

_____ 4

_____ 5

_____ 6

_____ 7

_____ 8

_____ 9

_____ 10

_____ 11

_____ 12

_____ 13

_____ 14

_____ 15

Put in order

Arrange these structures in the order in which blood passes through them—assume the blood is about to leave the right atrium.

Aorta

Aortic semilunar valve

Left atrium

Left ventricle

Lungs

Mitral valve

Pulmonary artery

Pulmonary semilunar valve

Right ventricle

Superior/inferior vena cava

Tissues of the body

Tricuspid valve

Fill-in

1. The flaplike lateral wall of each atrium is called the __?__.
2. The __?__ valve is also known as the mitral valve or left AV valve.
3. The right AV valve is also known as the __?__ valve.
4. The aortic semilunar valve has __?__ pocketlike flaps of tissue.
5. The __?__ are fibrous structures that prevent the cuspid valves from prolapsing (bending backwards).
6. One-way flow of blood from the right ventricle is ensured by the presence of the __?__ valve.
7. Mitral valve prolapse, which is abnormal, may allow blood to enter the __?__ during contraction of the left ventricle.
8. The small cardiac vein and right coronary artery can be found along the right __?__ sulcus.
9. The great cardiac vein and anterior interventricular artery can be found along the anterior __?__ sulcus.
10. The __?__ is a muscular wall between the left and right ventricles.
11. The myocardium of the __?__ ventricle is thicker than the other.
12. The wall of the aorta is thicker/thinner? than the wall of the superior vena cava.
13. The __?__ are beamlike processes of the inner face of the myocardium.
14. The "point" of the heart is called the __?__.
15. In the sheep heart, the right atrium is __?__ to the right ventricle.

LAB EXERCISE 27

Electrical Activity of the Heart

In Lab Exercise 26, you learned some of the major anatomical features of the heart. This exercise challenges you to take a step beyond the basics and learn about the electrical nature of the heart. As part of your study, you will demonstrate one of the most often used clinical tools for assessing the health of the heart, **electrocardiography.**

This exercise calls for any of several electrocardiographs commonly available in educational laboratories.

Before You Begin

☐ Read the appropriate chapter in your textbook.

☐ Set your learning goals. When you finish this exercise, you should be able to:
- o describe the conduction system of the heart
- o demonstrate electrocardiography
- o identify the features of a typical electrocardiogram and explain their significance

☐ Prepare your materials:
- o electrocardiograph apparatus or computer system with ECG recording kit
- o alcohol swabs
- o ECG electrodes, electrode tape, and gel
- o table or cot
- o metric (mm) ruler
- o PHYSIOLOGY INTERACTIVE LAB SIMULATIONS (PH.I.L.S.) 4.0 and computer system (optional)

☐ **Carefully** read the directions and safety tips for this exercise before starting any procedure.

SAFETY FIRST!

The apparatus used in this exercise is powered by electricity, so appropriate cautions should be taken to avoid electrical hazards. Students with known heart problems may volunteer as subjects in the study of resting ECGs but should not participate in activities involving exercise or other stressful situations.

A. Electrocardiography

The **cardiac cycle** is the pattern of physiological events exhibited during each beat of the heart. One approach is to describe the cardiac cycle as an atrial, then ventricular, **systole** (contraction) and **diastole** (relaxation). In other words, the atria contract and relax, and the ventricles contract and relax, producing one pumping cycle of the heart. During the cardiac cycle, a variety of physiological characteristics of the heart and pumped blood measurably change. For example, the **blood pressure** within each chamber changes with the volume of the chamber and the strength of contraction of the myocardium. Certain *heart sounds* are heard as the valves snap shut and portions of the myocardium contract. The electrical properties of the heart change as action potentials travel through the myocardium. It is this electrical aspect of the cardiac cycle that is of interest in this activity.

As you know, all muscle requires *excitation* of the cell membrane coupled to *contraction* involving filaments within the cell. The excitation, taking the form of action potentials, is an electrical event in which the polarity of electrical potential across the cell membranes changes. Because so much myocardial tissue depolarizes at more or less the same time, a great deal of current flows around the heart tissue, and the depolarization is easily detected from the outside of the body. Around the beginning of the twentieth century, a German scientist named Einthoven was one of the first to use a recording voltmeter to measure this electrical activity of the heart and to describe the changes found in a typical cardiac cycle. This technique is called electrocardiography.

The **electrocardiogram (ECG** or **EKG)** is a wave-form graph produced by a recording voltmeter called an **electrocardiograph.** The graph is a line, on a paper or monitor, that rises and falls with changes in voltage (electric potential) between two points. Figure 27.1*a* shows the voltage changes seen in one typical cardiac cycle. For convenience, each wave is given a name:

- o **P wave**—This wave represents depolarization of the atrial walls, which precedes atrial systole.

- o **QRS complex**—This set of waves results from the more or less concurrent ventricular depolarization (preceding

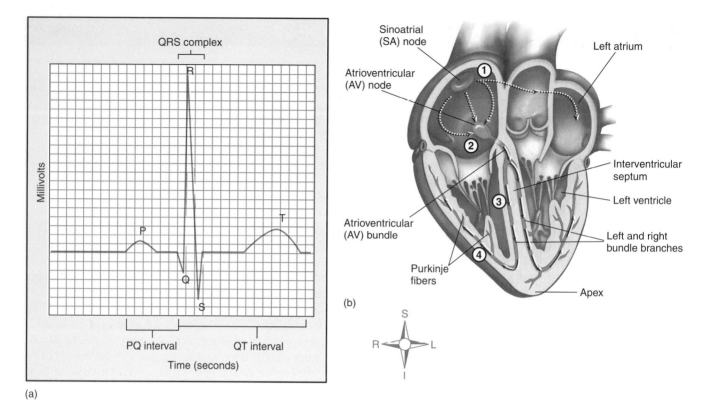

Figure 27.1　(*a*) Idealized ECG pattern in one cardiac cycle. (*b*) Anatomy of the conduction system of the heart.

ventricular systole) and atrial repolarization (signaling the onset of atrial diastole).

○　**T wave**—This wave represents ventricular repolarization, which precedes ventricular diastole.

The length of time during waves and between waves indicates the efficiency of conduction through the myocardium. Lengthened ECG portions are often evidence of *blocked conduction.* To understand the significance of conduction blockage, we must first outline the structure and function of the conduction system:

○　**Sinoatrial (SA) node**—This specialized section of cardiac muscle in the upper lateral wall of the right atrium generates its own action potentials at a faster rate than surrounding muscle, so it acts as a **pacemaker** for the rest of the heart (fig. 27.1*b*).

○　**Atrioventricular (AV) node**—A pacemaker in the lower right atrium subordinate to the SA node, it normally generates action potentials in response to potentials that travel to it through the atrial walls from the SA node.

○　**AV bundle** (*of His*)—This bundle of cardiac muscles in the interventricular septum is specialized to conduct action potentials rapidly. The *left* and *right AV bundle branches* spread out as **Purkinje fibers** through the lateral ventricular walls.

Action potentials generated by the SA node travel across the atrial myocardium, triggering atrial systole. The AV node then picks up the signal and generates action potentials that travel rapidly through the AV bundle and its branches to stimulate the ventricular myocardium, resulting in ventricular systole.

The **PQ interval** (sometimes termed the *PR interval*) is the time required for a signal to pass from the SA node through the atrial myocardium (depolarizing it) to the AV node and through its branches. The **QT interval** represents the time needed for complete depolarization and recovery of the ventricular myocardium.

Einthoven proposed three basic arrangements, or **leads,** of the ECG electrodes. Each lead is a set of one positive voltmeter electrode and one negative voltmeter electrode. These *standard leads,* or *appendicular leads,* are

○　**lead I**—one electrode on the left arm, one on the right arm

○　**lead II**—one electrode on the right arm, one on the left leg

○　**lead III**—one electrode on the left arm, one on the left leg

The electrode positions are shown in figure 27.2. The appendicular leads form what is called *Einthoven's triangle,* diagrammed in figure 27.3. In the diagram, the heart is shown at the peak of the R wave, when the ventricles are

Figure 27.2 Placement of ECG electrodes for the three standard (appendicular) leads.

partly depolarized. As you can see, an electric current flows from the positive area of the undepolarized region to the negative depolarized region. This difference in charge is detected by the ECG and is graphed as the R wave. Since the direction of current flow is close to the angle of lead II, the R wave of lead II is larger than in lead III and lead I.

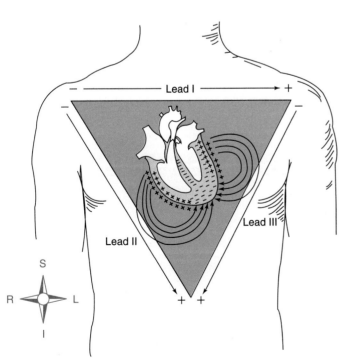

Figure 27.3 Einthoven's triangle.

Demonstrate the resting ECG:

☐ **1** Set up an electrocardiograph apparatus as directed by the instructions that accompany it. Set PAPER SPEED or TRACE VELOCITY to 25 mm/sec. (2.5 cm/sec.), the standard rate for ECG recording. Calibrate the SENSITIVITY or GAIN to 1 cm = 1 mV (see the instructions for your machine).

☐ **2** Have the (rested) subject lie on a cot or table or sit comfortably in a chair. Expose both ankles and wrists and clean a small area on each with alcohol pads. Put a *small* drop of electrode gel on each ECG electrode and place each on a different limb. Gently rub each electrode around to evenly distribute the gel; then anchor it with perforated rubber electrode tape pulled just tight enough to hold it in place.

☐ **3** Attach the limb wires to the appropriate electrode:
  ○ LA to the left arm electrode
  ○ RA to the right arm electrode
  ○ LL to the left leg electrode
  ○ RL to the right leg (for grounding the system)

☐ **4** Record lead I for about 20 cardiac cycles; then do likewise for leads II and III. Be sure to mark the part of the strip chart where you switched leads. If possible, record an ECG of every student in your group.

☐ **5** Using your lead II results, label a P wave, QRS complex, and T wave. If the P wave is absent and the heart rate is low, **SA node block** may be indicated. This could result from *ischemia* (reduced myocardial circulation) or *infarction* (myocardial damage).

27-4

□ **6** Using lead II results, measure these values:
- Heart rate (in beats per minute)—Remember that the paper rate is 25 mm/sec. (each mm = 0.04 sec.). **Tachycardia** is the condition of a heart rate above 100 beats/min and may be due to excitement (stress), high body temperature, or toxicity. **Bradycardia** is the condition of lowered heart rate (below 60 beats/min.). It is normal in conditioned athletes because they have a high stroke volume, or it may result from excessive vagal stimulation.
- PQ (PR) interval—If greater than 0.2 sec., a **first-degree heart block** could be present. This may result from inflammation of the AV bundle, which slows down conduction of a signal to the ventricles. If PQ is 0.25 to 0.45 sec. (and some P waves are not followed by QRS complexes), **second-degree heart block** may be present. This could result from AV node damage or excessive vagal stimulation. If the P wave seems independent of the QRS complex, with a P rate of about 100 per minute and a QRS of less than 40 per minute, **complete heart block** may be present.
- Your instructor may ask you to make other measurements and consult your textbook or a reference book for guidance in interpreting your results.

B. Deviations from the Normal ECG

A variety of factors may result in ECG readings far different than those expected. For example, the conditions described in activity A, step 6, may produce variations in the resting heart rate. In this activity, you are challenged to demonstrate some factors that alter the standard ECG pattern.

□ **1** Set up the ECG demonstration as described in activity A. Set the device to record lead II.

□ **2** Once a normal ECG pattern is observed, have the subject breathe deeply in and out several times. What happens? What consequences might this have in a clinical setting?

□ **3** Have the subject tense the muscles of the upper arm. What happens? What do you think causes this result? What implications does this observation have in clinical ECG applications?

□ **4** Remove the ECG cables, leaving the electrodes in place. Have the subject exercise for 2 to 5 minutes; then quickly reconnect the ECG cables. With the subject again in resting position, record 15 or 20 cardiac cycles. Continue to sample 15 to 20 cycles every minute for the next 10 minutes. Record the heart rate for each sample. How do you explain your results? Does exercise alter the features of each ECG cycle? Why or why not?

□ **5** What other factors might alter the ECG of the subject? As a group, discuss your hypotheses and, under the supervision of your instructor, perform experiments to test them (as time permits). Write a report of your experiment(s) and attach it to Lab Report 27. Be sure to include a statement of your hypothesis, an outline of your experimental design (methods), your results, and your interpretation of the results.

C. Computerized Electrocardiography

Many computer-based ECG systems are available for student use. If your lab has a computerized ECG setup, set it up according to the manual that accompanies it or the instruction screen on the computer. Most computerized ECG systems have their own instructions on the computer screen. You may also follow the steps given in activities A and B. In any case, collect the information asked for in Lab Report 27 and attach printouts of your ECG data/graphs to the lab report.

Most computerized ECG systems allow you to measure ECG intervals on screen automatically or with very little trouble. You may usually also save your ECG data and analyze them at a later time.

SAFETY FIRST!

Observe the safety precautions given earlier in this exercise and in your computerized system's manual.

202

D. The Augmented Leads

Many computerized ECG programs are designed to print out the results of the standard (appendicular) leads as well as three **augmented leads.** You may recall that Einthoven's triangle diagrammatically represents the direction of electric potential measurements using the three standard electrode arrangements (leads I, II, and III) (fig. 27.4).

Einthoven described a number of mathematical relationships among the data produced by the three leads. For example, knowing data from any two leads is sufficient to mathematically reconstruct a data line for the third lead. Information from combinations of two leads can be manipulated to produce the three augmented, or *attenuated,* leads. These are not *bipolar* leads or electrode placements but are the result of combining two negative electrodes and measuring it against one positive electrode. Thus, three more leads are produced:

aV$_R$—the LA and LL are combined to form the negative electrode; RA is the positive electrode
aV$_L$—the RA and LL electrodes form the negative electrode; LA is the positive electrode
aV$_F$—the RA and LA electrodes form the negative pole and LL the positive pole

Using Einthoven's triangle concept, the augmented leads are each shifted 90° from the standard leads. See figure 27.4.

Current clinical practice calls for the recording of 12 leads in all. So far, we have discussed the 3 standard leads and the 3 augmented leads. The remaining 6 leads are the **chest,** or **precordial,** leads. The chest leads use RA, LA, and LL combined to form the negative pole and each of 6 chest electrodes to form the positive poles. The chest leads are named V$_1$ through V$_6$.

What is the advantage of chest leads?

Hint . . . The computerized ECG system manual gives troubleshooting tips, as well as additional experiments.

(a)

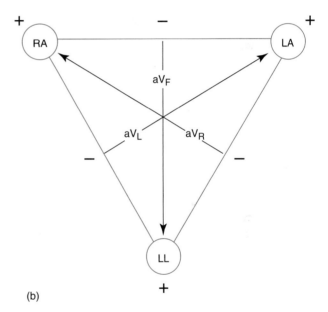
(b)

Figure 27.4 (*a*) Einthoven's triangle. (*b*) Einthoven's triangle superimposed on the three augmented leads.

Cardiac Dysrhythmia

The normal rhythm of the heartbeat (see fig. 27.1*a*) can be disturbed by a variety of conditions that affect the heart muscle or its conduction system (see fig. 27.1*b*).

Heart block occurs when the conduction system is "blocked" by damage to the myocardium. In *complete heart block,* P waves do not match up at all with the QRS complexes. In *bundle branch block,* the QRS complexes are prolonged (fig. 27.5).

Bradycardia is slow heart rhythm (less than 50 beats/min.); **tachycardia** is very rapid heart rhythm (more than 100 beats/min.). Either can be caused by normal conditions, such as deep sleep or exercise, or by abnormal conditions, such as the effects of drugs, blood loss (shock), fever, and damaged heart tissue.

Premature contractions (*extrasystoles*) are contractions that happen before the next expected contraction in series of heart beats. Occasional premature contractions are not a cause for alarm in otherwise healthy people. However, frequent premature contractions in a person with heart disease can reduce the pumping efficiency of the heart.

Frequent premature contractions can lead to **fibrillation**— a condition in which cardiac muscle fibers contract out of step with each other. *Atrial fibrillation,* or *A fib,* is fibrillation of the atrial myocardium. *Ventricular fibrillation,* or *V fib,* is fibrillation of the ventricular myocardium.

Complete heart block (P waves and QRS complexes are not coordinated)

Premature ventricular contraction (PVC) (no P waves precede PVCs)

Bundle branch block

Atrial fibrillation (no clear P waves and rapid QRS complexes)

Ventricular fibrillation (no P, QRS, or T waves)

Figure 27.5   Cardiac dysrhythmia (examples).

Advanced Scientific Inquiry
PHYSIOLOGY INTERACTIVE LAB SIMULATION (PH.I.L.S. 4.0)

If available, use the computerized laboratory simulation Ph.I.L.S. 4.0 to simulate one or more of the following activities that explore cardiac function:

Frog Heart Function

- *Thyroid and Chemical Effects (simulation #21)*
- *Refractory Period of the Heart (simulation #22)*
- *Starling's Law of the Heart (simulation #23)*
- *Heart Block (simulation #24)*

ECG and Heart Function

- *ECG and Exercise (simulation #25)*
- *The Meaning of Heart Sounds (simulation #26)*
- *ECG and Finger Pulse (simulation #27)*
- *Electrical Axis of the Heart (simulation #28)*
- *ECG and Heart Block (simulation #29)*
- *Abnormal ECGs (simulation #30)*

For each of the activities you perform, print the lab report according to the directions given in the activity and attach a copy of all sheets to your lab report (or as your instructor directs).

LAB REPORT 27

Electrical Activity of the Heart

Name _____

Date _____ Section _____

Test Condition	Heart Rate	PQ (PR) Interval	Other Observations	Interpretation
Resting				
Deep breathing				
Tensed arm				
Postexercise: 0 min.				
1 min.				
2 min.				
3 min.				
4 min.				
5 min.				
6 min.				
7 min.				
8 min.				
9 min.				
10 min.				

Staple a copy of your ECG graphs to this report. Attach reports of any additional ECG experiments you did.

Use this grid to graph the heart rates observed after exercise. Be sure to label your graph completely and accurately.

Put in order

_____ 1

_____ 2

_____ 3

_____ 4

_____ 5

_____ 6

_____ 7

Fill-in

_____ 1

_____ 2

_____ 3

_____ 4

_____ 5

Put in order

Arrange these heart structures in the order through which electrical signals are conducted in a single cardiac cycle.

atrial myocardium

AV bundle

AV bundle branches

AV node

Purkinje fibers

SA node

ventricular myocardium

Fill-in

1. Relaxation of a heart chamber is called __?__.
2. Contraction of a heart chamber is called __?__.
3. The portion of the ECG that represents ventricular repolarization is the __?__.
4. Leads I, II, and III together are called the __?__ leads, or appendicular leads.
5. __?__ is the condition of elevated heart rate.

The Pulse and Blood Pressure

The heart pushes blood from the left ventricle into the aorta with great force, as you already know. The wall of the aorta and its branches stretch a little with the sudden burst of blood during each ventricular ejection. Burst after burst after burst of blood ejected from the heart produce bulge after bulge after bulge in the arterial wall. These **pressure waves,** known as **pulse waves** or simply as the **pulse,** occur at more or less the same rhythm as the cardiac cycle. The pulse waves can be detected in clinical situations and used to determine many things about cardiovascular function. The rhythm and strength of the pulse in distal arteries can indicate how efficiently blood is being pumped into the circulation. The pressure of the waves can be measured indirectly and used to give information about overall arterial **blood pressure.** This exercise challenges you to find and measure some characteristics of blood pressure waves and use that skill to find some factors that affect cardiovascular function.

Before You Begin

☐ Read the appropriate chapter in your textbook.

☐ Set your learning goals. When you finish this exercise, you should be able to:
   o explain the concept of the pressure wave, or pulse
   o demonstrate detection of the pulse in a human subject
   o use a sphygmomanometer to measure blood pressure
   o list some factors that affect pulse and blood pressure

☐ Prepare your materials:
   o sphygmomanometer
   o stethoscope
   o alcohol swabs
   o watch or timer
   o PHYSIOLOGY INTERACTIVE LAB SIMULATIONS (PH.I.L.S.) 4.0 and computer system (optional)

☐ **Carefully** read the directions and safety tips for this exercise before starting any procedure.

A. The Pulse

As blood is ejected from the left ventricle during each cardiac cycle, it is subjected to great pressure. Since fluids are unable to be compressed, this great pressure pushes the compliant walls of the arteries outward. Between ejections, the blood pressure is lower, and the arterial walls bounce back to their original diameter. Because this occurs with each cardiac cycle, multiple waves of high pressure move into the arteries. These pressure waves are also called the pulse.

The rate and strength of the pulse can be measured easily with no special equipment. By simply pressing on the skin and pushing an artery against bone, one can feel the pressure waves. Counting the number of waves per minute gives the *pulse rate,* and the pulse strength (pressure) can be assessed by receptors in your fingertips. Determine the pulse rate of a resting subject by placing your second and third fingers on the anterior surface of the wrist, over the point where the *radial artery* passes over the distal end of the radius. (Don't use your thumb, or you may feel your own pulse waves.) Using a watch, count the number of waves in 15 seconds. Multiply your result by 4 to determine the pulse rate in waves per second. Try one or more of the other pulse points identified in figure 28.1. Are the pulse waves of equal pressure at each pulse point? Why or why not?

B. Blood Pressure

Biologists often measure the pressure of body fluids in units called **millimeters of mercury (mm Hg).** This unit is based on a measuring device called a *manometer,* which is an inverted tube of liquid mercury. As something presses against the mercury reservoir at the bottom of the tube, the column of mercury in the tube rises higher. The more pressure with which the mercury is pushed, the higher the column. Thus, the more millimeters the mercury rises (mm Hg), the higher the pressure (fig. 28.2*a*). A **sphygmomanometer** is a manometer with an air cuff attached to the reservoir; it is used to indirectly measure the changing blood pressures associated with the pulse.

☐ **1** Use a sphygmomanometer to measure the blood pressure of a resting subject. First, clean the ear pieces of a stethoscope with an alcohol swab and place them in your ears. Then wrap the sphygmomanometer air cuff around the subject's left upper arm, just above the elbow.

☐ **2** Place the stethoscope sensor in the position shown in figure 28.2*b* and listen to the pulse in the *brachial artery.* Close the valve of the sphygmomanometer's air bulb, and pump the cuff to a pressure of about 175 mm Hg.

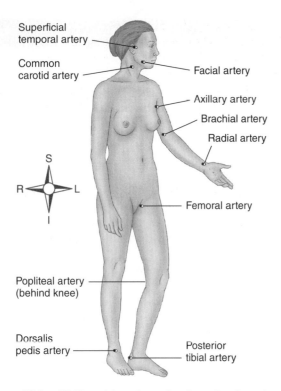

Figure 28.1 Well-positioned arteries for pulse detection.

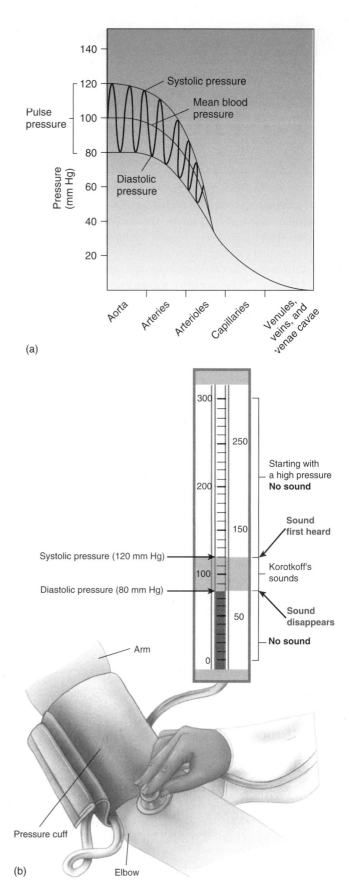

(a)

□ **3** Still listening to the artery and watching the mercury
column, open the valve slightly and slowly release air.

o As you decrease the pressure, you will suddenly
notice loud, tapping **Korotkoff sounds** in the
artery. These sounds are caused by the pulsing of
blood pressure against the wall of the artery when
the pressure in the cuff equals the pressure in the
artery. Note the pressure at which the sounds
begin.

o Continue to decrease pressure until the Korotkoff
sounds disappear, also noting that point on the
pressure scale. The sounds disappear when the
cuff pressure drops below the pressure in
the artery.

o The point at which the sounds are first heard is
the **systolic pressure,** the pressure of the pulse
wave's crest (fig. 28.2a). The sounds disappear
when the **diastolic pressure,** the blood pressure
between pulse waves, is reached. Blood
pressure is usually expressed as systolic
pressure over diastolic pressure. The blood
pressure shown here is about 120 over 80, or
120 mm Hg/80 mm Hg. This may be considered
"normal" or "high normal" if it is a person's
average blood pressure.

o **Auscultation,** from the Latin "listening", refers to
any procedure in which you listen to body sounds.
You have just performed an *auscultatory method* of
blood pressure determinination.

(b)

Figure 28.2   (*a*) The changing pressures associated with the
pulse wave. (*b*) Using a sphygmomanometer.

C. Variations in Pulse and Pressure

Determine the pulse rate and blood pressure in the situations described. Compare the results with each other and with the results for a resting subject. How do you account for the variations?

□ 1 Test a variety of body positions: lying down (immediately, then after 5 minutes) and standing (immediately, then after 5 minutes).

□ 2 Test a subject after exercise: immediately after 2 to 5 minutes of exercise, then once per minute for the next 10 minutes of recovery. Compare your results with those for the postexercise ECG (Lab Exercise 27).

Advanced Scientific Inquiry
PHYSIOLOGY INTERACTIVE LAB SIMULATION (PH.I.L.S. 4.0)

If available, use the computerized laboratory simulation PH.I.L.S. 4.0 to simulate one or more of the following activities that explore circulatory function:

- *Cooling and Peripheral Blood Flow (simulation #31)*
- *Blood Pressure and Gravity (simulation #32)*
- *Blood Pressure and Body Position (simulation #33)*

For each of the activities you perform, print the lab report according to the directions given in the activity and attach a copy of all sheets to your lab report (or as your instructor directs).

The Pulse and
Blood Pressure

Name _____

Date _____ Section _____

Test Condition	Pulse Rate	Pulse Strength	Systolic Blood Pressure	Diastolic Blood Pressure	Pulse Pressure (Systolic Minus Diastolic)
Reading: *radial*					
Resting: _____					
Resting: _____					
Lying down: 0 min.					
5 min.					
Standing: 0 min.					
5 min.					
After exercise 0 min.					
1 min.					
2 min.					
3 min.					
4 min.					
5 min.					
6 min.					
7 min.					
8 min.					
9 min.					
10 min.					

The next page is titled "Interpretation." In the space provided, write your interpretation of the results reported in this table. For example, tell why the strength of the resting pulse rate may have varied depending on where it was detected. Look for similarities and differences from row to row (how does pulse rate in different body positions compare?). Look for relationships from column to column (does pulse strength relate to pulse pressure?). Use the questions given in the exercise itself for further guidance.

28-6

Interpretation

LAB EXERCISE 29

The Circulatory Pathway

Blood pumped from the right ventricle of the heart travels to and through the lungs before returning to the left atrium. This pathway is termed the **pulmonary circulation,** or the *pulmonary loop*. Blood pumped from the left ventricle travels to and through the other organs and tissues of the body. This pathway is termed the **systemic circulation** (*systemic loop*). These two pathways, together with the heart, form the basic structure of the cardiovascular system. In this exercise, you are challenged to learn some of the major blood vessels that form the circulatory pathways of the human body.

Before You Begin

☐ Read the appropriate chapter in your textbook.

☐ Set your learning goals. When you finish this exercise, you should be able to:
  o locate major veins and arteries in models and charts
  o describe the circulatory pathways to and from major body regions

☐ Prepare your materials:
  o model of the human body (showing major vessels)
  o charts of human circulatory routes

☐ **Carefully** read the directions and safety tips for this exercise before starting any procedure.

A. Arteries

Arteries are blood vessels that conduct blood away from the heart and toward tissues (fig. 29.1). In the pulmonary circulation, **pulmonary arteries** conduct deoxygenated blood to the lungs. In the systemic circulation, the **aorta** and its branches conduct oxygenated blood toward the systemic tissues. Small arteries are usually called **arterioles.** Arterioles conduct blood into a network of even smaller vessels, or **capillaries.** On a chart, then in a model of the human body, locate the major human arteries listed. As you do so, note how they form routes to major regions of the body.

☐ **1** Find these portions of the aorta:
  o **ascending aorta**—the first portion, before the aorta bends inferiorly

  o **aortic arch**—the bend of the aorta, just superior to the heart
  o **descending aorta**—the remainder of the aorta
  o **thoracic aorta**—the portion of the descending aorta that is within the thorax
  o **abdominal aorta**—the portion of the descending aorta that is within the abdomen

☐ **2** Locate these arteries of the head and neck:
  o **brachiocephalic**—the first branch from the aorta
  o **right common carotid**—the medial branch of the brachiocephalic artery
  o **right subclavian**—the lateral branch of the brachiocephalic artery
  o **left common carotid, left subclavian**—the second and third branches, respectively, of the aorta
  o **internal carotid, external carotid**—the branches of the common carotid arteries
  o **vertebral**—the branches of the subclavians
  o **basilar**—the artery formed by fusion of vertebral arteries
  o **circle of Willis**—the circular system of arteries around the brain's base, formed by branches of the basilar artery and the internal carotids

☐ **3** Identify these arteries of the upper limb:
  o **axillary artery**—a continuation of the subclavian artery inferior to the clavicle
  o **brachial**—a continuation of the axillary artery in the upper arm
  o **ulnar**—the medial branch of the brachial artery
  o **radial**—the lateral branch of the brachial artery

☐ **4** Identify these branches of the descending aorta:
  o **intercostal**—branches from the aorta to the intercostal muscles
  o **phrenic**—branches from the aorta to the diaphragm
  o **celiac**—the branch of the aorta to the stomach, pancreas, and liver
  o **superior mesenteric**—the branch of the aorta to the small intestine and proximal large intestine
  o **inferior mesenteric**—the branch of the aorta to the distal regions of the large intestine
  o **renal**—branches from the aorta to the kidneys
  o **suprarenal**—branches from the aorta to the adrenal glands

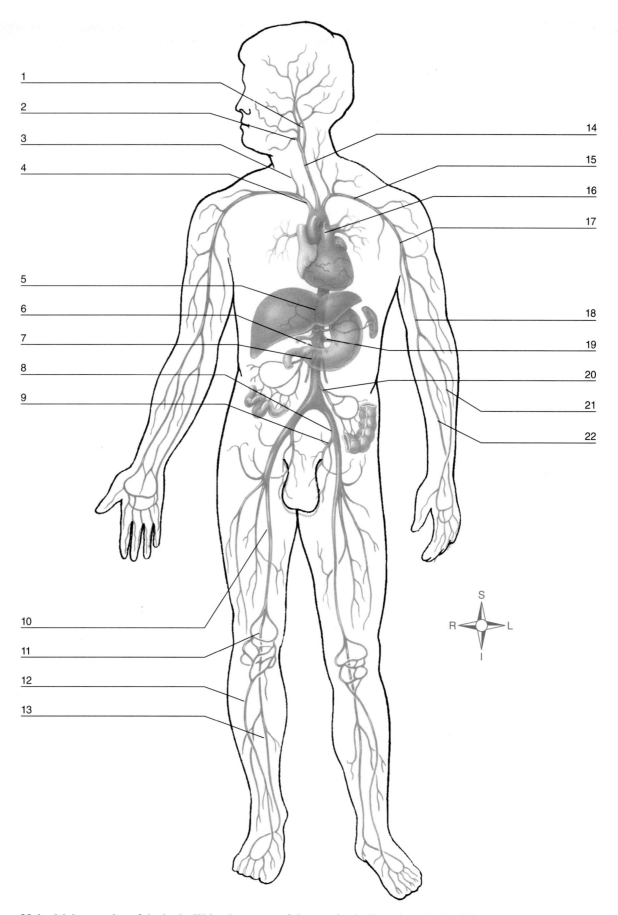

Figure 29.1 Major arteries of the body. Write the names of the arteries indicated on the label lines.

- o **testicular, ovarian**—branches from the aorta to the gonads
- o **common iliac**—branch from the inferior end of the descending aorta toward a leg

☐ **5**   Identify these arteries of the pelvis and lower limbs:
- o **internal iliac**—the medial branch of the common iliacs
- o **external iliac**—the lateral branch of the common iliac
- o **femoral**—a continuation of the external iliac artery in the thigh
- o **popliteal**—a continuation of the femoral artery in the popliteal area (posterior knee)
- o **anterior tibial**—the anterior branch of the popliteal artery
- o **posterior tibial**—the posterior branch of the popliteal artery

B. Veins

Veins are blood vessels that conduct blood toward the heart (fig. 29.2). In the pulmonary circulation, the **pulmonary veins** return oxygenated blood from the lungs. In the systemic circulation, the **superior vena cava** returns deoxygenated blood from the head, neck, thorax, and arms. The **inferior vena cava** returns deoxygenated blood from the rest of the systemic loop. **Venules** are small veins. In charts, then in a model, locate the listed major veins. As with the arteries, note how they form routes from major body regions.

☐ **1**   Identify these veins of the head and neck:
- o **brachiocephalic**—the medial branch into the superior vena cava
- o **subclavian**—the lateral branch into the brachiocephalic vein
- o **internal jugular**—the medial branch into the brachiocephalic vein
- o **external jugular**—the external vein of the neck that returns blood to the subclavian vein

☐ **2**   Identify these veins of the upper limb and thorax:
- o **axillary**—the medial branch into the subclavian vein
- o **basilic vein**—the superficial vein that empties into the axillary vein
- o **brachial**—the upper arm vein that continues into the axillary region as the axillary vein
- o **cephalic**—the lateral, superficial branch into the subclavian vein
- o **azygos**—the unpaired branch into the posterior aspect of the superior vena cava
- o **hemiazygos, accessory hemiazygos**—two sets of multiple veins that empty into the azygos

- o **intercostal**—veins that empty into the azygos vein (right) and hemiazygos or accessory hemiazygos veins (left)

☐ **3**   Identify these tributaries of the inferior vena cava:
- o **hepatic**—the vein from the liver to the inferior vena cava
- o **renal**—the vein from the kidney to the inferior vena cava
- o **testicular, ovarian**—the vein from the gonad to the inferior vena cava
- o **common iliac**—two branches that fuse to become the inferior vena cava
- o **internal iliac**—the medial branch of the common iliac (in the pelvis)

☐ **4**   A **portal circulation** is a set of vessels that begins and ends with capillary networks. In other words, blood is returned to a second set of capillaries before being returned to the heart. The **hepatic portal system** is an important part of the venous systemic circulation (fig. 29.3). It returns blood from digestive organs to the liver, rather than directly to the heart. Identify these hepatic portal vessels:
- o **hepatic portal**—from the veins of abdominal organs to the liver
- o **superior mesenteric**—from the small intestine to the hepatic portal vein
- o **inferior mesenteric**—from the large intestine, joining the splenic vein to the hepatic portal vein
- o **splenic**—from the spleen to the hepatic portal vein
- o **gastroepiploic**—from the stomach, joining the splenic vein to the hepatic portal vein

What physiological purpose can be served by the presence of the hepatic portal system?

☐ **5**   Identify these veins of the lower limbs:
- o **external iliac**—the lateral branch into the common iliac vein (in the pelvis)
- o **femoral**—the major lateral branch into the external iliac vein
- o **great saphenous**—the major medial, superficial branch into the external iliac vein
- o **popliteal**—the posterior branch into the femoral vein, on the posterior of the knee
- o **small saphenous**—the lateral, superficial branch into the popliteal vein, lateral to the tibia
- o **anterior tibial**—branch into the popliteal vein, on the anterior aspect of the tibia
- o **posterior tibial**—the posterior to the tibia, draining into the popliteal vein

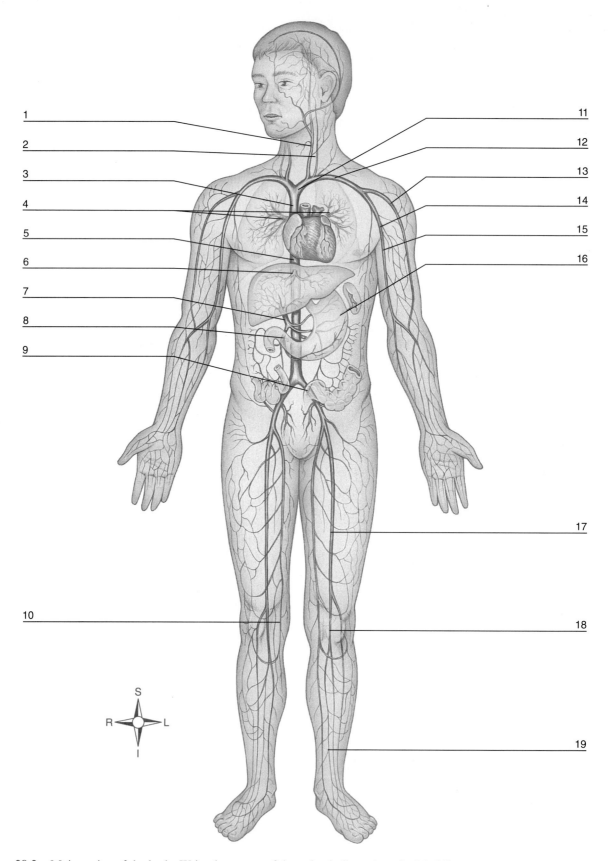

1 _____

2 _____

3 _____

4 _____

5 _____

6 _____

7 _____

8 _____

9 _____

10 _____

11 _____

12 _____

13 _____

14 _____

15 _____

16 _____

17 _____

18 _____

19 _____

Figure 29.2   Major veins of the body. Write the names of the veins indicated on the label lines.

3

4

5

1

6

2

S

R L

I

Figure 29.3 The hepatic portal system. Write the names of the vessels indicated on the label lines.

Angiography

Angiography is a radiographic (x-ray) imaging method used to visualize blood vessels. A *radiopaque* dye, a water-soluble material that absorbs x-rays so that it casts a heavy shadow on the film, is injected near the vessels of interest. When used to visualize the heart and nearby vessels, the resulting image is called an *angiocardiogram*. When used to visualize veins, the resulting image is called a *venogram*. When angiography is used to image arteries, the result is called an *arteriogram*.

Angiography is used to visualize the structure of blood vessels to determine cardiovascular health. Some abnormalities that may be seen in an angiogram include

- blood vessel injuries
- *aneurysms,* abnormal local distensions of a vessel wall
- blockage, perhaps from *atherosclerosis* (buildup of cholesterol and other matter on the inside of a vessel wall)

Figure 29.4 is an arteriogram in which radiopaque dye has been injected into the femoral artery. Try to identify each artery labeled in the figure.

1 _____

2 _____

3 _____

4 _____

5 _____

Figure 29.4   Arteriogram. Try to identify the indicated arteries.

LAB REPORT 29

The Circulatory Pathway

Name _____

Date _____ Section _____

Figure 29.1 *(from p. 216)*

_____ 1
_____ 2
_____ 3
_____ 4
_____ 5
_____ 6
_____ 7
_____ 8
_____ 9
_____ 10
_____ 11
_____ 12
_____ 13
_____ 14
_____ 15
_____ 16
_____ 17
_____ 18
_____ 19
_____ 20
_____ 21
_____ 22

Put in order

left ventricle 1
_____ 2
_____ 3
_____ 4
_____ 5
_____ 6
_____ 7
_____ 8
_____ 9
_____ 10
_____ 11
_____ 12
_____ 13
_____ 14
_____ 15
_____ 16
_____ 17
_____ 18

Identify

_____ 1
_____ 2
_____ 3
_____ 4
_____ 5
_____ 6
_____ 7
_____ 8

Put in order

Arrange these structures in the order in which blood passes through them in a circuit from the heart to the foot and back—the first one is done for you.

abdominal aorta

anterior tibial artery

anterior tibial vein

aortic arch

ascending aorta

common iliac artery

common iliac vein

external iliac artery

external iliac vein

femoral artery

femoral vein

foot

inferior vena cava

left ventricle

popliteal artery

popliteal vein

right atrium

thoracic aorta

Identify

1. Portion of the aorta in the abdomen
2. Continuation of the subclavian artery inferior to the clavicle
3. Artery that supplies the diaphragm
4. Vein that drains the liver
5. Vessel that conducts blood from digestive organs to the liver
6. Major medial tributary of the external iliac vein
7. Unpaired vein that drains into the posterior aspect of the superior vena cava
8. First major branch of the aorta

Figure 29.2 *(from p. 218)*

_____ 1

_____ 2

_____ 3

_____ 4

_____ 5

_____ 6

_____ 7

_____ 8

_____ 9

_____ 10

_____ 11

_____ 12

_____ 13

_____ 14

_____ 15

_____ 16

_____ 17

_____ 18

_____ 19

Figure 29.3 *(from p. 219)*

_____ 1

_____ 2

_____ 3

_____ 4

_____ 5

_____ 6

Figure 29.4 *(from p. 220)*

_____ 1

_____ 2

_____ 3

_____ 4

_____ 5

Vessel	Tissues Supplied or Drained
Brachiocephalic artery	
Left common carotid artery	
Left subclavian artery	
Celiac artery	
Phrenic arteries	
Superior mesenteric artery	
Inferior mesenteric artery	
Renal artery	
Suprarenal artery	
External iliac artery	
Internal iliac artery	
Hepatic vein	
External iliac vein	
Internal iliac vein	
Renal vein	
Testicular vein	
Ovarian vein	
Phrenic vein	
Superior mesenteric vein	
Inferior mesenteric vein	
Gastroepiploic vein	
Cystic vein	

The Lymphatic System and Immunity

The **lymphatic system** has several primary functions: it participates in maintaining extracellular fluid balance, it absorbs fats and other substances from the digestive tract, and it functions in the **immune system** defense of the body. To do this, the lymphatic system is composed of a number of vessels and other structures. This exercise presents some of the basics of lymphatic anatomy and physiology.

Before You Begin

☐ Read the appropriate chapter in your textbook.

☐ Set your learning goals. When you finish this exercise, you should be able to:
   o describe the major organs of the lymphatic system and find them in a chart
   o identify the features of a lymph node
   o briefly state the function of lymph nodes

☐ Prepare your materials:
   o charts or models of the lymphatic system

☐ **Carefully** read the directions and safety tips for this exercise before starting any procedure.

A. Overview of the Lymphatic System

Use a chart or model of the lymphatic system to find these gross features:

☐ **1** Locate some of the **lymphatic vessels.** These vessels are similar to veins in the structure of their walls and the presence of many valves. Lymph vessels collect **lymph** from interstitial spaces in the regions served and conduct it toward the blood circulation. Lymph, or *lymphatic fluid,* is an excess from interstitial areas. The **right lymphatic duct** is a large collecting vessel that receives lymph from the superior right quadrant of the body. The **thoracic duct** similarly conducts lymph received from the rest of the body. Each duct empties into a subclavian vein.

☐ **2** **Lymph nodes** are small, round organs located at irregular intervals in the network of lymph vessels.

Like all lymph organs, they contain **lymphatic tissue** mainly composed of lymphocytes (a category of white blood cell). They are distributed unevenly, with major aggregations of nodes in the neck (**cervical nodes),** armpit (**axillary nodes),** and groin (**inguinal nodes).**

☐ **3** The **tonsils** are lymph organs that surround the openings of the mouth, nose, and throat into the lower digestive and respiratory tracts. The **palatine tonsils** are on each lateral side of the opening of the mouth into the throat. The **pharyngeal tonsils** are at the top of the throat (*pharynx*) near the posterior of the nasal cavity. The **lingual tonsils** are on the posterior of the tongue's base.

☐ **4** The **spleen** and **thymus** should already be familiar to you. The spleen is located in the upper left abdominal cavity and the thymus is located in the anterior mediastinum.

Hint . . . A microscopic study of thymic tissue is offered in Lab Exercise 23.

B. The Lymph Node

There are numerous lymph nodes throughout the lymphatic system, as figure 30.1 shows. Lymph usually passes through one or more lymph nodes on its path toward the blood circulation. The materials in lymph may stimulate lymphocyte development (an immune response) or may be destroyed by *macrophages* in the node. Identify these structures associated with the lymph node in a chart or model:

☐ **1** **Afferent lymph vessels** conduct lymph into a lymph node, whereas an **efferent lymph vessel** conducts lymph out of each node.

☐ **2** Within the node, lymph tissue often forms dense masses called **lymph nodules.** Each nodule contains a **germinal center** composed of rapidly dividing lymphocytes. Surrounding each node is a space through which lymph circulates; this space is called a **lymph sinus.** The sinuses also contain macrophages.

☐ **3** The fibrous **capsule** functions as the outer boundary and support structure for each lymph node. The inner surface of the capsule has inward extensions that separate the nodules from one another.

☐ **4** Complete figure 30.2 to review what you've learned about the tonsils and lymph nodes.

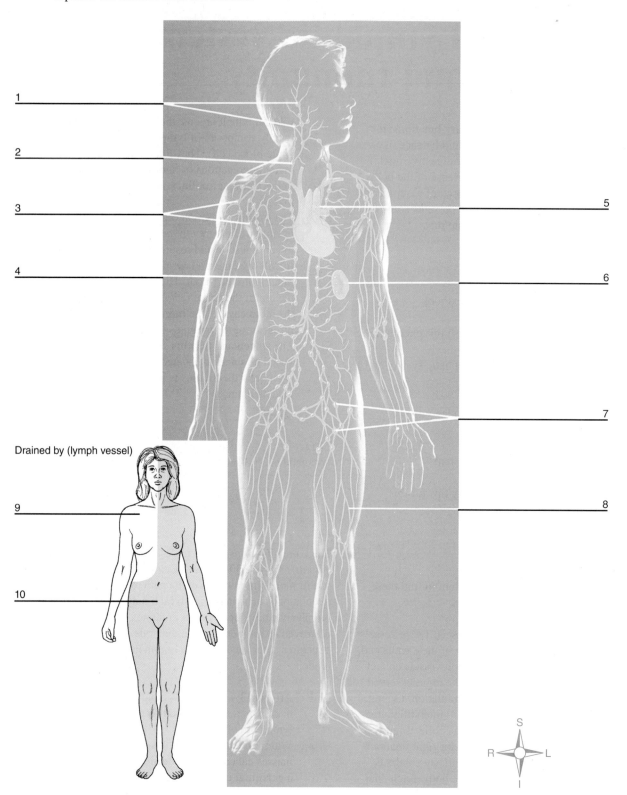

1 _____

2 _____

3 _____

4 _____

5 _____

6 _____

7 _____

8 _____

Drained by (lymph vessel)

9 _____

10 _____

Figure 30.1 Label the indicated lymphatic structures. The inset shows the drainage regions served by the two main lymphatic ducts. Name them.

The Tonsils

PALATINE TONSILS 1
PHARYNGEAL TONSILS 2
LINGUAL TONSILS 3

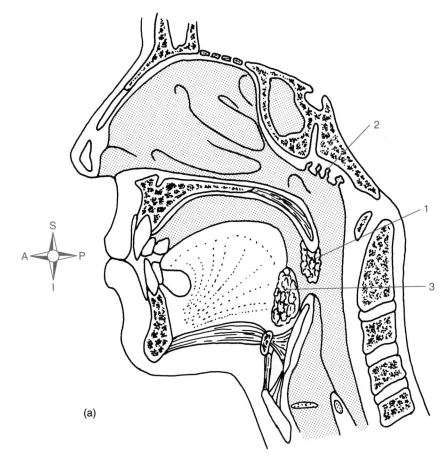

(a)

The Lymph Node

AFFERENT VESSEL 1
EFFERENT VESSEL 2
LYMPH NODULE 3
   GERMINAL CENTER 4
LYMPH SINUS 5
CAPSULE 6
ARTERY 7
VEIN 8

(b)

Figure 30.2 With colored pens or pencils color (*a*) the tonsils and (*b*) the features of a lymph node.

Clinical Applications

Lymphocytes may be involved in two major types of responses to the presence of potentially threatening foreign substances (fig. 30.3). In **antibody-mediated immunity,** *B lymphocytes* may produce antibodies that react with specific antigens on an invading cell or molecule. In **cell-mediated immunity,** *T lymphocytes* secrete **cytokines** that signal other immune responses and often destroy antigen-containing cells directly.

The antibody-antigen reactions associated with immune responses have been used by clinical biologists for years. Several examples follow. Think about each example and answer the questions posed.

1. Biologists often use antibodies to test for the presence of certain antigens in a particular substance. In Lab Exercises 24 and 25, you witnessed demonstrations of this technique. For each of the following antigens, indicate the substance tested (blood, urine, etc.) and give a brief summary of the antibody-antigen reaction involved. (Use a separate sheet for your answers.)

   Human chorionic gonadotropin:
   Luteinizing hormone:
   A and B antigens:
   D antigen:

2. *Rheumatoid arthritis* is an inflammatory disease affecting joint tissues. This disease is known to have an *autoimmune* component. Autoimmunity is an immune response inappropriately directed toward normal self antigens. Many persons with rheumatoid arthritis have *rheumatoid factor* (*RF*) present in their blood. RF is an abnormal antibody. Not all victims of the disease have RF in their plasma, and RF is known to be present in conditions other than rheumatoid arthritis. Based on this information and your previous study, outline a simple test to detect the presence of RF in plasma. If your test is used to screen for rheumatoid arthritis, does a positive result (i.e., RF is present) mean that the person definitely has the disease? Why or why not? (Use a separate sheet for your answers.)

Figure 30.3   Immune reactions. The antigen (*a*) can react with receptors (*b*) located on the surface of T lymphocytes or with antibodies (*c*) produced by B lymphocytes during the immune response of the body.

LAB REPORT 30

The Lymphatic System and Immunity

Name _____

Date _____ Section _____

Figure 30.1 *(from p. 224)*

_____ 1

_____ 2

_____ 3

_____ 4

_____ 5

_____ 6

_____ 7

_____ 8

_____ 9

_____ 10

Multiple choice

_____ 1

_____ 2

_____ 3

_____ 4

_____ 5

_____ 6

_____ 7

Multiple choice

1. Lymphatic tissue mainly contains
   a. lymphocytes.
   b. erythrocytes.
   c. monocytes.

2. The large lymphatic organ found in the left hypochondriac region is the
   a. thymus.
   b. palatine tonsil.
   c. spleen.
   d. lingual tonsil.
   e. inguinal lymph node.

3. The lymph from tissues of the lower limbs is conducted by the
   a. right lymphatic duct.
   b. thoracic duct.

4. Lymph nodes
   a. provide a site for lymphocyte reproduction.
   b. anchor lymphatic vessels.
   c. provide a site for phagocytosis of foreign cells or debris.
   d. *a* and *b*.
   e. *a* and *c*.

5. Lymph vessels that conduct lymph into a lymph node are termed
   a. efferent lymphatic vessels.
   b. efferent arterioles.
   c. afferent arterioles.
   d. afferent lymphatic vessels.
   e. none of the above.

6. The lymphatic system
   a. maintains extracellular fluid balance.
   b. collects excess fluid from the interstitial spaces.
   c. defends the body against foreign cells and molecules.
   d. *a* and *c*.
   e. *a*, *b*, and *c*.

7. Lymph drains directly into the blood at
   a. the right subclavian vein.
   b. the left subclavian vein.
   c. the subclavian arteries.
   d. *a* and *b*.
   e. none of the above.

Lymphatic Structure	Location	Function(s)
Lymphatic capillaries		
Lymphatic vessels		
Right lymphatic duct		
Thoracic duct		
Lymph node		
Palatine tonsils		
Pharyngeal tonsils		
Lingual tonsils		
Spleen		
Thymus		

Sketch

Draw a cross section of a lymph node and label its parts.

Respiratory Structures

This exercise is the first of two that deal with the **respiratory system.** In the activities presented here, you will explore the essential structure of the *respiratory tract.* Lab Exercise 32, on the other hand, will deal with some aspects of respiratory function.

The respiratory tract consists of the nose and nasal cavity, pharynx, larynx, trachea, bronchi, and lungs. During **ventilation,** a process that includes *respiratory cycles* of **inspiration,** then **expiration,** air moves in and out of the respiratory tract. During inspiration, air moves through the tract to the lungs, and it moves out during expiration. While in the lungs, air exchanges oxygen and carbon dioxide in a process called **gas exchange.** Keep these important functions of the system in mind as you explore the gross and microscopic structures presented in this exercise.

Before You Begin

☐ Read the appropriate chapter in your textbook.

☐ Set your learning goals. When you finish this exercise, you should be able to:
  o describe the major organs of the respiratory system
  o locate respiratory structures on a chart or model
  o demonstrate the structure of respiratory organs in a preserved specimen
  o trace the movement of air through the respiratory tract
  o locate structural features of the trachea in a microscopic specimen
  o locate structural features of lung tissue in a microscopic specimen

☐ Prepare your materials:
  o chart or model of the human respiratory system
  o dissection tools and large tray
  o preserved specimen: *sheep pluck*
  o microscope
  o prepared microslide: *mammalian trachea c.s.*
  o prepared microslide: *human lung c.s.*

☐ **Carefully** read the directions and safety tips for this exercise before starting any procedure.

A. Gross Anatomy

Using a chart or model of the respiratory tract (perhaps a dissectible human head and torso), locate the following structural features of the respiratory tract (fig. 31.1).

☐ 1 **Nose** is a term often reserved for the external cartilage and bone forming the anterior wall of the **nasal cavity** but may refer to the entire cavity. Identify these nasal features:
  o **nasal septum**—the wall dividing the nasal cavity into right and left portions
  o **nares** (sing., **naris**)—*external nares,* or the *nostrils,* and the *internal nares,* or openings from the posterior nasal cavity into the pharynx
  o **hard palate**—the floor of the nasal cavity
  o **conchae** (sing., **concha**)—three curved ridges extending from each lateral wall of the cavity: the *superior nasal concha, middle nasal concha,* and *inferior nasal concha*
  o **paranasal sinuses**—air-filled, mucus-lined spaces in the skull that communicate with the nasal cavity: *maxillary, frontal, ethmoid,* and *sphenoid sinuses*
  o **nasolacrimal duct**—the tube that conducts tears from the conjunctiva over the eyes and empties into the nasal cavity
☐ 2 Identify these features of the **pharynx,** or throat:
  o **nasopharynx**—the upper of three regions of the pharynx, this one posterior to the nasal cavity; the **uvula** is an extension of the floor of the nasopharynx, or **soft palate,** and serves as the inferior boundary landmark of the nasopharynx; the *pharyngeal tonsils* are on the posterior wall, and the lateral walls have the inferior opening of the *Eustachian tube* from the middle ear
  o **oropharynx**—the middle portion of the pharynx, posterior to the oral cavity; the *palatine* and *lingual tonsils* are found here
  o **laryngopharynx**—the most inferior portion of the pharynx, posterior to the larynx
☐ 3 Identify these features of the voice box, or **larynx:**
  o *laryngeal cartilages*—nine pieces of cartilage (six pairs and three single pieces) that fit together

Figure 31.1 Midsagittal section of the head, showing respiratory structures. Label the indicated structures.

to form the wall of the larynx: **thyroid, cricoid, epiglottis, cuneiforms** (pair), **corniculates** (pair), and **arytenoids** (pair)

- o **vestibular folds (false vocal cords)**—the superior of two pairs of ligaments stretched across the lateral portions of the larynx cavity
- o **vocal folds (true vocal cords)**—the inferior of two pairs of laryngeal ligaments

☐ **4** The **trachea,** or windpipe, conducts air between the larynx above and the bronchi below. It has C-shaped cartilage rings embedded in its anterior and lateral walls for support. At its inferior end, the trachea divides into the left and right **primary bronchi,** each of which enters a lung at a **hilum.**

☐ **5** The **lungs** are divided, on a gross level, into **lobes,** which can be subdivided into **lobules.** The left lung has a *superior lobe* and an *inferior lobe.* The right lung has a *superior, a middle,* and an *inferior lobe.* The lungs are found in the *pleural cavities* in the left and right portions of the thorax. A serous membrane called the **pleura** covers each lung (*visceral pleura*), then folds back to line the pleural cavity (*parietal pleura*). The thin space between the visceral and parietal layers is filled with serous **pleural fluid.**

☐ **6** Once in a lung, each primary bronchus branches to form a **bronchial tree.** The successive levels of bronchial branching include
- o **secondary bronchi**—each serving a separate lobe
- o **tertiary bronchi**—each a branch of a secondary bronchus; each tertiary bronchus serves a lobule
- o **bronchioles**—several levels of small branches of the airway within each lobule; near the end of the branching, **terminal bronchioles** give rise to **respiratory bronchioles,** which in turn subdivide to form **alveolar ducts** that end as groups of air sacs called **alveoli;** the smaller bronchioles are not represented in many models but can be found in some charts and figures

☐ **7** The **diaphragm** is a sheet of skeletal muscle just inferior to both lungs. The diaphragm contracts during inspiration to reduce pressure inside the thorax, drawing air into the respiratory tract.

☐ **8** Complete figure 31.2.

B. Dissection

This activity calls for individual dissection or a demonstrated dissection of a *sheep pluck.* The pluck is a portion of the respiratory tract and other organs removed from an animal when it is slaughtered. Notice that the heart and diaphragm are still intact. This is a good opportunity to reinforce your understanding of their anatomical relationship to one another.

SAFETY FIRST!

Remember to observe the standard precautions for dissecting a preserved specimen.

☐ **1** Identify the larynx, trachea, primary bronchi, lungs, heart, and diaphragm.

☐ **2** Examine the lungs. Can you distinguish the lobes? How many lobes are in each lung? Is it the same number as in the human?

☐ **3** Use a scissors to cut along the dorsal wall of the respiratory tract to open the larynx and trachea. What features can you identify? Continue your cut through a primary bronchus and into the bronchial tree. How many levels of branching can you see? Can you name the different branches?

C. Microscopic Anatomy

SAFETY FIRST!

Use the microscope and prepared slides with caution. Don't forget the electrical hazards and possibility of cuts from broken slides.

☐ **1** Examine a prepared cross section of a mammalian trachea. Identify these features:
- o Lining—It is composed of *pseudostratified columnar epithelium.* Locate *cilia* and *goblet cells* if you can.
- o Outer wall—It is composed of *smooth muscle* and assorted connective tissues. Locate portions of the C-shaped cartilage supports in the wall of the trachea. What type of cartilage is it?

☐ **2** Examine a prepared cross section of human lung tissue. Identify this feature:
- o **Alveolus**—It is an air sac that communicates with a respiratory bronchiole. Try to locate the *alveolar epithelium,* the simple squamous epithelium forming the wall. Is any of the connective tissue between alveoli visible in your specimen?

☐ **3** Try to locate the *pulmonary capillaries* that surround the alveoli. In a cross section, they appear as circles of simple squamous epithelium (and sometimes have blood cells within them).

Respiratory Anatomy

NASAL CAVITY 1
NASOPHARYNX 2
OROPHARYNX 3
LARYNGOPHARYNX 4
LARYNX 5
TRACHEA 6
PRIMARY BRONCHI 7
SECONDARY BRONCHI 8
TERTIARY BRONCHI 9
BRONCHIOLE 10
TERMINAL BRONCHIOLE 11

RESPIRATORY BRONCHIOLE 12
ALVEOLAR DUCT 13
ALVEOLUS 14

LOBES (Shade lightly)
RIGHT SUPERIOR LOBE 15
RIGHT MIDDLE LOBE 16
RIGHT INFERIOR LOBE 17
LEFT SUPERIOR LOBE 18
LEFT INFERIOR LOBE 19

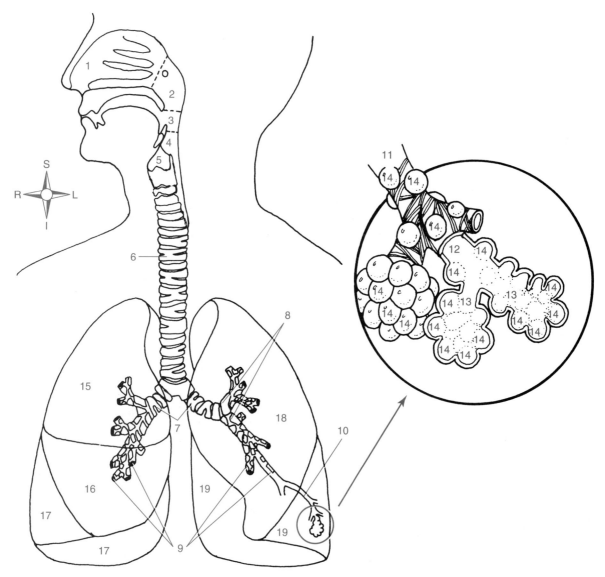

Figure 31.2   Use colored pens or pencils to color these respiratory structures and their labels.

Respiratory Structures

Name _____

Date _____ Section _____

Use this table as a checklist for your study of the human model and the sheep pluck. Use your textbook for help with the section that asks for functions.

Structure	Human	Pluck	Functions(s)
Nasal cavity	☐	☐	
Nasal septum	☐	☐	
Nares: internal, external	☐	☐	
Hard palate	☐	☐	
Soft palate	☐	☐	
Conchae: superior, middle, inferior	☐	☐	
Paranasal sinuses: maxillary, frontal, ethmoid, sphenoid	☐	☐	
Nasopharynx	☐	☐	
Oropharynx	☐	☐	
Laryngopharynx	☐	☐	
Laryngeal cartilages	☐	☐	
Vestibular folds	☐	☐	
Vocal folds	☐	☐	
Trachea	☐	☐	
Primary bronchi	☐	☐	
Secondary bronchi	☐	☐	
Tertiary bronchi	☐	☐	
Lungs: lobes, lobules	☐	☐	
Pleurae: visceral, parietal	☐	☐	
Bronchioles	☐	☐	
Terminal bronchioles	☐	☐	
Respiratory bronchioles	☐	☐	
Alveolar ducts	☐	☐	
Alveoli	☐	☐	

Specimen: *trachea c.s.*
        Total magnification: _____

Specimen: *human lung c.s.*
        Total magnification: _____

Sketch and label the sheep pluck specimen:

Figure 31.1 *(from p. 230)*

_____ 1

_____ 2

_____ 3

_____ 4

_____ 5

_____ 6

_____ 7

_____ 8

_____ 9

_____ 10

_____ 11

_____ 12

_____ 13

_____ 14

_____ 15

_____ 16

_____ 17

_____ 18

_____ 19

_____ 20

_____ 21

_____ 22

_____ 23

Put in order

_____ 1

_____ 2

_____ 3

_____ 4

_____ 5

_____ 6

_____ 7

_____ 8

_____ 9

Put in order

Arrange these structures in the order through which air passes during inspiration.

larynx

nasal cavity

pharynx

primary bronchi

respiratory bronchioles

secondary bronchi

terminal bronchioles

tertiary bronchi

trachea

LAB EXERCISE 32

Pulmonary Volumes and Capacities

Spirometry is the use of instrumentation to determine the basic **pulmonary volumes** and **pulmonary capacities.** Pulmonary volumes, or *lung volumes,* are the amounts of air that are moved in and out of the respiratory tract during different phases of the respiratory cycle. Pulmonary capacities are combinations of pulmonary volumes, as you will see. Spirometry is important in many applications: research, clinical situations, and exercise training.

In this exercise, you will use a **spirometer** to measure pulmonary volumes and capacities.

Before You Begin

☐ Read the appropriate chapter in your textbook.

☐ Set your learning goals. When you finish this exercise, you should be able to:
  ○ define the volumes and capacities of the respiratory tract and explain their functional significance
  ○ demonstrate the use of a spirometer to determine pulmonary volumes and capacities
  ○ understand the effects of exercise on pulmonary volumes and capacities

☐ Prepare your materials:
  ○ wet or dry (handheld) spirometer
  ○ one-way valve and mouthpiece (for wet spirometer) or disposable mouthpieces (dry)
  ○ series of bowls of disinfectant
  ○ noseclip
  ○ BIOHAZARD bag
  ○ stopwatch or timer
  ○ Computerized spirometry system (optional)
  ○ computer system (if computerized version is used)
  ○ PHYSIOLOGY INTERACTIVE LAB SIMULATIONS (PH.I.L.S.) 4.0 and computer system (optional)

☐ **Carefully** read the directions and safety tips for this exercise before starting any procedure.

A. Pulmonary Volumes and Capacities

Inspiration, or **inhalation,** is one of two major phases of the respiratory cycle. **Expiration,** or **exhalation,** is the other. The movement of air into and out of the respiratory tract during these phases is termed **ventilation** and is critical to the normal functioning of the entire body. As mentioned in the introduction, a spirometer is an instrument that measures the volume of air moved during the different phases of the respiratory cycle. This activity calls for the use of a *wet spirometer,* which is a vessel inverted in a container of water. A breathing tube communicates with the inverted vessel, which rises in the water as it fills with air. The rise of the inverted vessel is proportional to the volume exhaled into the instrument. A recording spirometer, as shown in figure 32.1, produces a wavy **spirogram.** You will probably use a spirometer that shows the rise on a dial or linear scale. You may instead use a handheld *dry spirometer,* which is a gauge that measures air volumes as you exhale into it.

Before using the spirometer, you should practice breathing normally while wearing a noseclip (or pinching your nostrils shut). The noseclip is needed to prevent the natural tendency to breathe through your nose, which invalidates the spirometry results.

Figure 32.1 A recording wet spirometer. A recording spirometer yields a wave-form spirogram, but you will probably use a nonrecording spirometer that shows changing pulmonary volume as movements on a dial or scale.

SAFETY FIRST!

A one-way valve and mouthpiece used with a wet spirometer prevents inhalation of potentially contaminated air from inside the spirometer's air passages. However, be certain that the valves and mouthpieces are thoroughly disinfected by soaking them in a series of alcohol baths before reusing them. If a handheld spirometer is used, be sure to use a *new* disposable mouthpiece *and* disinfect the spirometer itself between uses (fig. 32.2).

Figure 32.2 Insert a disposable mouthpiece over the stem of a hand-held spirometer. Dispose of used mouthpieces in a biohazard container.

☐ **1** **Tidal volume (TV)** is the volume of air moved into or out of the respiratory tract during normal breathing. Measure your TV as follows.

If using a one-way valve:
  ○ Place your mouth on the tube.
  ○ Inhale normally.
  ○ Exhale normally.
  ○ Read the result and reset the apparatus to zero.

If not using a one-way value:
  ○ Inhale normally.
  ○ Place your mouth on the tube.
  ○ Exhale normally.
  ○ Read the result and reset the apparatus to zero.

Hint . . . Pulmonary volumes are reported in either milliliters (mL) or liters (L). Remember that milliliters are a thousandth the size of liters. Therefore, a typical TV might be reported as 500 mL or 0.5 L. Be sure to indicate which units you are using when reporting your results.

• Perform each test three times and average your results.

• Typical volumes and capacities are given in figure 32.3 and in figure 32.4, shown in the box on the facing page.

☐ **2** **Expiratory reserve volume (ERV)** is the volume of air that can be expired forcefully after a normal inspiration. Measure the ERV as follows.

If using a one-way valve:
  ○ place your mouth on the tube
  ○ inhale normally
  ○ exhale forcefully (to the maximum)

Pulmonary Volumes and Capacities

Volume	Definition	Typical Value
Tidal volume (TV)	Volume moved into or out of the respiratory tract during a normal respiratory cycle	0.5 L
Inspiratory reserve volume (IRV)	Maximum volume that can be moved into the respiratory tract after a normal inspiration	3.0 L
Expiratory reserve volume (ERV)	Maximum volume that can be moved out of the respiratory tract after a normal expiration	1.1 L
Residual volume (RV)	Volume remaining in the respiratory tract after maximum expiration	1.2 L

Capacity	Formula	Typical Value
Vital capacity (VC)	TV + IRV + ERV	4.6 L
Inspiratory capacity (IC)	TV + IRV	3.5 L
Functional residual capacity (FRC)	ERV + RV	2.3 L
Total lung capacity (TLC)	TV + IRV + ERV + RV	5.8 L

Figure 32.3 Wave diagram of pulmonary volumes and capacities.

Figure 32.4 Balloon diagram of pulmonary volumes and capacities.

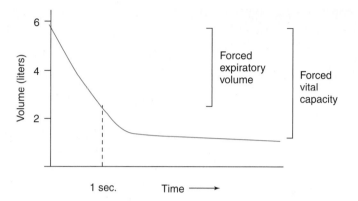

Figure 32.5 Spirogram of the forced expiratory vital capacity (FVC) test. In normal subjects, about 80% of the total forced vital capacity is exhaled during the first second of forced rapid expiration.

     o Read the result and subtract your TV—the difference is ERV; reset the apparatus to zero.
If not using a one-way valve:
     o Inhale and exhale normally.
     o At the end of the normal expiration, place your mouth on the tube and exhale forcefully.
     o Read the result as ERV and reset the apparatus to zero.

☐ **3** **Vital capacity (VC)** is the volume of air that can be expired after a forceful inspiration. Determine your vital capacity as follows.
If using a one-way valve:
     o Place your mouth on the tube.
     o Inhale forcefully (to the maximum).
     o Exhale forcefully (to the maximum).
     o Read the VC value and reset the apparatus to zero.
If not using a one-way valve:
     o Inhale forcefully (to the maximum).
     o Place your mouth on the tube.
     o Exhale forcefully (to the maximum).
     o Read the VC value and reset the apparatus to zero.

☐ **4** **Inspiratory reserve volume (IRV)** is the volume of air that can be forcefully inspired after a normal inspiration. Because your spirometer is not really set up to do this, determine the IRV by subtracting TV and ERV from the vital capacity:

$$IRV = VC - (TV + ERV)$$

☐ **5** Analyze and interpret your results. Compare TV, ERV, VC, and IRV with the values given in the box. How do you account for any differences?

B. Forced Expiratory Vital Capacity

The **forced expiratory vital capacity (FVC)** test is useful in determining the nature of pulmonary conditions. In the FVC test, the subject maximally inhales, then maximally exhales as rapidly as possible. The event is timed and a reading taken at exactly 1 second after the beginning of expiration. The 1-second volume is called the **forced expiratory volume** at 1 second, or **FEV_1**. The total amount expired, however long it takes, is the FVC (fig. 32.5). If the FEV_1 is much less than 80% of the FVC, then airway obstruction is suspected. For example, airway constriction in asthma produces a low FEV_1/FVC percentage. The increased resistance to airflow in constricted airways prevents air from being expired rapidly.

The spirometry apparatus used here is not the best for measuring FEV_1 values. However, a rough estimate can be made by following the procedure described.

☐ **1** Designate a SUBJECT, a TIMER, and a RECORDER in your lab team. As the SUBJECT prepares to exhale into the spirometer tube, the TIMER stands ready to call "time" at exactly 1 second. The RECORDER will mark the spirometer scale at (or take note of) the volume indicated when "time" was called.

☐ **2** *If a one-way valve is used:*
     o Have the SUBJECT place his/her mouth on the tube.
     o Have the SUBJECT inhale maximally, then exhale as forcefully and rapidly as possible until the maximum expiration is reached.
If a one-way valve is not used:
     o Have the SUBJECT inhale maximally.
     o Have the SUBJECT place his/her mouth on the tube.
     o Have the SUBJECT exhale as forcefully and rapidly as possible until maximum expiration is reached.

☐ **3** The TIMER calls "time" at 1 second after the beginning of exhalation, and the RECORDER marks or makes note of the spirometer volume at exactly that point. This volume is FEV_1. Record the volume at the end of forceful expiration. This volume is FVC.

☐ **4**   Determine the FEV$_1$/FVC percentage by using this equation:

$$FEV_1/FVC \% = (FEV_1 \div FVC) \times 100$$

Is the result 80% or above? Try the test on a subject with a known obstructive pulmonary condition. What are your results?

C. The Effects of Exercise

Try repeating activity A with a person after exercise. After determining resting values, have the SUBJECT exercise heavily for 2 to 5 minutes. Immediately after exercise has stopped, measure the pulmonary volumes again. How have they changed? How do you account for the changes? Try measuring the volume after about 20 minutes of rest. Explain your results.

D. Computerized Spirometry

Many computer-based spirometry systems are available for student use. If your lab has a computerized spirometry setup, set it up according to the manual that accompanies it or the instruction screen on the computer. Most computerized physiology systems have their own instructions for specific activities on the computer screen. You may instead follow the instructions in the previous Activities of this Exercise. In any case, collect the information requested in Lab Report 32 (or by your instructor) and attach printouts of your computerized experiments to your lab report.

Advanced Scientific Inquiry
Physiology Interactive Lab Simulation (Ph.I.L.S. 4.0)

If available, use the computerized laboratory simulation Ph.I.L.S. 4.0 to simulate one or more of the following activities that explore digestive function:

- *Altering Body Position (simulation #37)*
- *Altering Airway Volume (simulation #38)*
- *Exercise Induced Changes (simulation #39)*
- *Deep Breathing and Cardiac Function (simulation #40)*

You may also want to try these activities that explore hemoglobin's oxygen-binding characteristics:

- *pH and Hb-oxygen Binding (simulation #34)*
- *DPG and Hb-oxygen Binding (simulation #35)*

For each of the activities you perform, print the lab report according to the directions given in the activity and attach a copy of all sheets to your lab report (or as your instructor directs).

Pulmonary Volumes and Capacities

Name _____

Date _____ Section _____

Resting:

Volume or Capacity	Expected Value[1]	Actual Value[2]	Difference	Interpretation
Tidal volume (TV)				
(Average)	0.5 L (500 mL)			
Expiratory reserve volume (ERV)				
(Average)	1.1 L (1,100 mL)			
Vital capacity (VC)				
(Average)	4.6 L (4,600 mL)			
Inspiratory reserve volume (IRV)	3.0 L (3,000 mL)			

1. Predicted values are general approximations. Factors such as body surface area, sex, age, and athletic conditioning affect predicted values. If a "predicted vital capacity chart" is available, use the figure from that rather than the figure given here.
2. Be sure to indicate units for each reported value (milliliters or liters).

Trial	FEV$_1$	FVC	FEV$_1$/FVC%	Interpretation
1				
2				
3				
4				
Best value				

After exercise (0 min.):

Volume or Capacity	Expected Value[1]	Actual Value	Difference	Interpretation
TV				
ERV				
VC				
IRV				

1. Use the results from the first table on the previous page as your expected results. Be sure to use the correct units of volume when reporting your values.

After exercise (20 min.):

Volume or Capacity	Expected Value[1]	Actual Value	Difference	Interpretation
TV				
ERV				
VC				
IRV				

1. Use the results from the first table on the previous page as your expected results. Be sure to use the correct units of volume when reporting your values.

Discuss your results (comparing resting values with values after exercise):

LAB EXERCISE 33

Digestive Structures

The **digestive tract** is a series of hollow organs through which food passes: mouth, pharynx, esophagus, stomach, small intestine, and large intestine. Each portion is specialized for one or more aspects of the three major functions of the **digestive system:** *secretion, digestion,* and *absorption.* Accessory organs, such as the **salivary glands, liver, gallbladder,** and **pancreas,** have ducts that lead into the digestive tract and thus support digestive function.

This exercise challenges you to explore the structure and some of the essential functions of the digestive system.

Before You Begin

☐ Read the appropriate chapter in your textbook.

☐ Set your learning goals. When you finish this exercise, you should be able to:
  o describe the structure of digestive organs and locate them in models and charts
  o identify the principal function of each major digestive organ
  o describe the basic histology of the gastrointestinal wall

☐ Prepare your materials:
  o model of the human torso (dissectible) or chart
  o human skull (with teeth intact)
  o model or specimen of a tooth section
  o microscope
  o prepared microslides: *stomach wall c.s.*
                          small intestine wall c.s.

☐ **Carefully** read the directions and safety tips for this exercise before starting any procedure.

A. The Wall of the Digestive Tract

The wall of the digestive tract forms a hollow tube continuous with the external environment. Food passes along the tube in a process usually termed **motility.** The wall of the tract is composed of four layers (tunics). They are

  o **mucosa**—the inner lining, consisting of mucous epithelium, connective tissue (*lamina propria*), and a thin layer of smooth muscle (*muscularis mucosa*)

  o **submucosa**—a thick layer of connective tissue, with blood vessels, nerves, and glands, just superficial to the mucosa

  o **muscularis**—layers of smooth muscle surrounding the submucosa, usually an inner *circular layer* and an outer *longitudinal layer* (the stomach also has a middle *oblique*) layer

  o **serosa (adventitia)**—the most superficial layer of the digestive wall, composed of connective tissue (adventitia) or epithelial tissue (serosa)

☐ 1 Locate the digestive tract in a dissectible model of the human torso. Some portions may have the wall cut in a cross section. If so, try to identify each of the four layers. How does each layer differ from organ to organ?

SAFETY FIRST!

Observe the usual precautions when using the microscope and prepared slides.

☐ 2 Obtain a prepared slide of a sectioned stomach wall and examine it with your microscope. Identify the four layers of the wall. Examine the mucosa. Notice that it is composed of simple columnar epithelium that forms deep **gastric pits.** They are lined with **surface mucous cells,** which you may recognize as goblet cells. Follow a pit's lumen to the base, or **gastric gland.** You may see more goblet cells, called **mucous neck cells.** Some of the very darkly stained cells are **chief cells.** The very pale cells are **parietal cells.** What is the function of each of these cells?

☐ 3 Obtain a prepared slide of a small intestine section. Identify the four layers of the wall. In a true cross section, you can't tell that the simple columnar epithelium forms large **circular folds,** but you may have noticed them in the model. Each circular fold has fingerlike projections, called **villi,** that are visible in your specimen.

B. The Upper Digestive Tract

Use a dissectible model or another aid to identify these digestive structures:

☐ **1** The **mouth,** or **oral cavity,** is the beginning of the one-way human digestive tract (fig. 33.1). Locate these features:

- ○ **lips**
- ○ **buccinator muscle**—the muscle that forms most of the cheek walls
- ○ **tongue**—a mass of muscle covered with mucous epithelium containing *taste buds;* the **frenulum** is a fold of tissue anchoring its underside to the floor of the mouth
- ○ **hard palate**—the anterior, bony portion of the mouth's roof
- ○ **soft palate**—the posterior, soft portion of the mouth's roof (it ends in the conelike **uvula**)
- ○ **palatine tonsils**—lymphatic structures on the posterior, lateral aspect of the mouth cavity

☐ **2** The 32 **teeth** of the normal adult are special features of the mouth cavity (fig. 33.2*b*). Identify these tooth

types in a skull or model (each type has a representative on each side, left and right, of each jaw, upper and lower):

- ○ **incisors**—*central* and *lateral* cutting teeth
- ○ **canine**—tearing teeth
- ○ **first** and **second premolars**—tearing and cutting teeth, each with two points, or *cusps*
- ○ **first, second,** and **third molars**—grinding teeth with three flattened cusps

Find each of these features on a sectioned tooth or tooth model:

- ○ **crown**—the superior portion of the tooth
- ○ **neck**—the narrow portion, below the crown
- ○ **root**—the portion of the tooth within the **alveoli,** or spaces, of the mandible and maxillae
- ○ **pulp cavity**—the central space of the tooth filled with **pulp,** which is connective tissue with nerves and vessels
- ○ **dentin**—the hard, calcified tissue that forms most of the tooth shell
- ○ **enamel**—the hard, nonliving, mineralized substance that covers the dentin of the crown
- ○ **gingiva**—the mucous epithelial covering of the *alveolar ridges* of the jaws
- ○ **periodontal membrane**—the fibrous lining of the alveolar walls (**periodontal ligaments** project from this membrane to the root of the tooth, keeping it in place)

Figure 33.1 Identify the indicated features of the mouth.

The Mouth and Teeth

LIPS 1
BUCCINATOR MUSCLE 2
TONGUE 3
HARD PALATE 4
SOFT PALATE 5
UVULA 6
PALATINE TONSILS 7
OROPHARYNX 8

TEETH

CENTRAL INCISOR 9
LATERAL INCISOR 10
CANINE 11
FIRST PREMOLAR 12
SECOND PREMOLAR 13
FIRST MOLAR 14
SECOND MOLAR 15
THIRD MOLAR 16

TOOTH FEATURES

CROWN 17
NECK 18
ROOT 19
PULP CAVITY 20
PULP 21
DENTIN 22
ENAMEL 23
GINGIVA 24
PERIODONTAL MEMBRANE 25
PERIODONTAL LIGAMENTS 26
ALVEOLAR RIDGE 27

 Figure 33.2 Use colored pens or pencils to color the labels and the features of (*a*) the mouth and (*b*) a typical tooth.

☐ **3** The **salivary glands** are exocrine glands whose ducts empty into the mouth cavity. **Saliva** produced by the glands is a watery solution containing mucus, mineral ions, and digestive enzymes. Identify these pairs of salivary glands in a model:

- **Parotid glands**—The largest pair, just anterior to each ear, each secretes a serous (watery) solution, which flows through a **parotid duct** leading to the cheek wall across from the second upper molar.
- **Submandibular glands**—Medial to the angle of the mandible, these glands produce mostly serous fluid, and some mucous fluid. Their ducts empty onto the mouth floor, just lateral to the frenulum.
- **Sublingual glands**—The smallest pair of salivary glands in the floor of the mouth, they produce mostly mucous fluid, which flows into the mouth through 10 or more small ducts in the mouth's floor.

☐ **4** The **pharynx,** or throat, is divided into three regions:
- **nasopharynx**
- **oropharynx**
- **laryngopharynx**

☐ **5** The **esophagus** is a long, muscular tube connecting the pharynx to the stomach. Upper and lower **esophageal sphincters** may be represented in the model you are using.

☐ **6** The **stomach** is an enlargement of the digestive tract just inferior to the diaphragm, at the end of the esophagus. Locate these stomach features in a model or chart:
- **cardiac opening**—the opening from the esophagus into the stomach region called the **cardiac region**
- **fundus**—the region of the stomach superior to the cardiac opening
- **body**—the main portion of the stomach, forming a **greater curvature** and a **lesser curvature**
- **pyloric region**—the portion of the stomach near the **pyloric opening** into the small intestine and the **pyloric sphincter** that surrounds the opening
- **rugae**—the large folds of the mucosa

C. The Lower Digestive Tract

☐ **1** The **small intestine** is a long, narrow tube that folds to fill a large portion of the abdominal cavity. Find these features of the small intestine:
- **duodenum**—the short (±25 cm), C-shaped beginning of the small intestine
- **jejunum**—the long (±2.5 m) middle section of the small intestine
- **ileum**—the very long (±3.5 m) end section of the small intestine
- **common bile duct**—the duct that empties into the duodenum

- **ileocecal junction**—the junction of the ileum with the large intestine, featuring an **ileocecal sphincter** and a one-way **ileocecal valve**

☐ **2** The **liver,** a large gland in the superior right abdominal cavity, produces **bile,** which is secreted into the duodenum. Two **hepatic ducts** conduct bile from the liver and fuse to form a **common hepatic duct.** The common hepatic duct joins the two-way **cystic duct** from the **gallbladder** (just inferior to the liver) to form the **common bile duct.** The common bile duct empties into the duodenum.

☐ **3** The **pancreas** is a gland, cradled in the C of the duodenum, that secretes digestive fluids. Exocrine pancreatic secretions flow through the **pancreatic duct,** which joins the common bile duct.

☐ **4** The **large intestine** includes these features:
- **cecum**—a blind sac extending inferiorly from the ileocecal junction; attached to the posterior wall of the cecum is a narrow, wormlike blind sac called the **appendix**
- **colon**—a long (±1.6 m) portion of the large intestine subdivided into the **ascending colon, transverse colon, descending colon,** and **sigmoid colon;** the longitudinal layer of the colon's muscularis forms three bands of muscle called the **teniae coli;** the wall of the colon forms **haustra,** or pouches; the exterior aspect of the colon wall has fatty attachments called **epiploic appendages.**
- **rectum**—a short, straight, muscular tube at the end of the sigmoid colon
- **anal canal**—the short (±2.5 cm), muscular canal, at the end of the rectum, that exits to the outside of the body by way of the **anus;** it features a smooth muscle **internal anal sphincter** at its superior end and a skeletal muscle **external anal sphincter** at its inferior end

☐ **5** The organs and inner wall of the abdominal cavity are covered with a serous membrane called the **peritoneum.** The portion that covers the organs is called the **visceral peritoneum,** and the portion that lines the cavity walls is called the **parietal peritoneum.** Extensions of the peritoneum, composed of a thin sheet of connective tissue sandwiched between two layers of serous membrane, support the visceral organs. These extensions are called **mesenteries.** Find these mesenteries in your model:
- **lesser omentum**—located between the lesser curvature of the stomach and the liver (and diaphragm)
- **greater omentum**—connecting the stomach's greater curvature to the transverse colon and posterior abdominal wall; it has a long, fatty, double fold forming an apronlike pocket called the **omental bursa** over the front of the viscera

☐ **6** Review what you've learned in activities B and C by completing figure 33.3.

The Digestive System

ORAL CAVITY 1
PAROTID GLAND 2
SUBMANDIBULAR GLAND 3
SUBLINGUAL GLAND 4
PHARYNX 5
ESOPHAGUS 6
LOWER ESOPHAGEAL
  SPHINCTER 7
FUNDUS 8
CARDIAC REGION 9
BODY OF STOMACH 10
PYLORIC REGION 11
PYLORIC SPHINCTER 12
DUODENUM 13
JEJUNUM 14
ILEUM 15
COMMON BILE DUCT 16
ILEOCECAL VALVE 17
LIVER 18
COMMON HEPATIC DUCT 19
CYSTIC DUCT 20
GALLBLADDER 21
PANCREAS 22
PANCREATIC DUCT 23
CECUM 24
APPENDIX 25
ASCENDING COLON 26
TRANSVERSE COLON 27
DESCENDING COLON 28
SIGMOID COLON 29
RECTUM 30
ANAL CANAL 31

 Figure 33.3   Use colored pens or pencils to color these organs of the digestive system. *Note:* many organs are shown separated from one another. In life, many digestive organs overlap one another when viewed from this perspective.

Fluoroscopy

Fluoroscopy is a type of radiographic imaging that uses a fluorescent screen rather than film. X-rays pass through the patient and are absorbed by the screen's coating. The energy is re-emitted by the coating as light energy. Dense tissues cast a dark shadow on the screen. Therefore, continuous radiation allows a continuous, moving image in a moving patient.

Fluoroscopy with the contrast medium **barium sulfate** is used to visualize portions of the digestive tract. A *barium swallow,* or **upper GI series,** images the stomach and portions of the esophagus and small intestine. A *barium enema* (**lower GI series**) images the rectum and colon. A **hernia** of the digestive tract, in which a portion of the tract is pushed out of a weak spot in the abdominopelvic wall,

can be detected this way. **Ulcers,** holes in the digestive wall, can also be assessed in this manner.

Because fluoroscopy allows observation of the body in motion, it can be used to see the muscular movements of the digestive tract as the barium passes through it. A person can push on the abdomen from the outside to see how digestive organs withstand pressure similar to the compression of abdominal muscles during heavy lifting.

Figure 33.4 consists of regular x-ray photographs taken of an upper and lower GI series. A fluoroscope image would be reversed (black on white—not the white on black seen here). Try to label the parts.

Figure 33.4   Radiographs of (*a*) a barium swallow, or upper GI, and (*b*) a barium enema, or lower GI. Note: Fluoroscope images appear black on white, the reverse of the white-on-black images shown here.

LAB REPORT 33

Digestive Structures

Name _____

Date _____ Section _____

Specimen: *stomach wall c.s.*
Total magnification: _____

Specimen: *small intestine wall c.s.*
Total magnification: _____

Structure	Model	Function(s)
Oral cavity	☐	
Buccinator muscle	☐	
Tongue	☐	
Palate: hard, soft	☐	
Uvula	☐	
Palatine tonsils	☐	
Incisors: central, lateral	☐	
Canine teeth	☐	
Premolars: first, second	☐	
Molars: first, second, third	☐	
Parotid glands	☐	
Submandibular glands	☐	
Sublingual glands	☐	
Pharynx	☐	
Esophagus	☐	
Cardiac opening, region	☐	
Fundus	☐	
Body of stomach	☐	
Stomach curvatures: greater, lesser	☐	
Pyloric opening, region	☐	

33-8

Structure	Model	Function(s)
Rugae	☐	
Duodenum	☐	
Jejunum	☐	
Ileum	☐	
Circular folds	☐	
Common bile duct	☐	
Ileocecal junction: valve, sphincter	☐	
Common hepatic duct	☐	
Cystic duct	☐	
Pancreas	☐	
Pancreatic duct	☐	
Cecum	☐	
Appendix	☐	
Ascending colon	☐	
Transverse colon	☐	
Descending colon	☐	
Rectum	☐	
Anal canal, anus	☐	
Lesser omentum	☐	
Greater omentum	☐	

Figure 33.1 *(from p. 244)*

_____ 1

_____ 2

_____ 3

_____ 4

_____ 5

_____ 6

_____ 7

Figure 33.4 *(from p. 248)*

_____ 1

_____ 2

_____ 3

_____ 4

_____ 5

_____ 6

_____ 7

_____ 8

_____ 9

Fill-in

_____ 1

_____ 2

_____ 3

_____ 4

_____ 5

_____ 6

_____ 7

Fill-in

1. The stomach's mucosa forms large folds called __?__.
2. The two ducts that exit the liver and join to form the common hepatic duct are called __?__ ducts.
3. The colon is divided into __?__ sections.
4. __?__ are fatty extensions of the colon's outer wall.
5. A substance called __?__ covers the dentin of the tooth's crown.
6. The __?__ is the portion of the pharynx posterior to the mouth cavity.
7. The __?__ sphincter prevents stomach contents from flowing back into the esophagus.

Enzymes and Digestion

Digestion is the breaking down of complex nutrients into simpler ones. Digestion involves mechanical processes that simply break the nutrients into smaller chunks and chemical processes that split large molecules into smaller molecules. In this exercise, you will look at examples of both types of digestive processes. As an example of a mechanical process, you will use **bile salts** secreted by the liver to break apart lipid droplets. You will use pancreatic **enzymes** to chemically break apart complex carbohydrate and lipid molecules. Enzymes are proteins that catalyze biochemical reactions.

Before You Begin

☐ Read the appropriate chapter in your textbook.

☐ Set your learning goals. When you finish this exercise, you should be able to:
  o describe the action of pancreatic enzymes on carbohydrates and lipids
  o explain the effects of temperature on enzyme activity
  o perform tests to determine the presence of starch and sugars
  o demonstrate the action of bile on lipids

☐ Prepare your materials:
  o distilled water
  o pancreatin powder (and spatula)
  o 1% starch solution
  o 1% maltose solution
  o Benedict's reagent
  o ice water bath (with test tube rack)
  o warm water bath (37°C, with tube rack)
  o hot plate
  o Lugol's reagent
  o disposable 1 mL measuring droppers
  o test tubes (approx. 10 mL) tube caps
  o test tube racks and tongs
  o wax pencils
  o beakers (500 mL)
  o bile salt powder (and spatula)
  o vegetable oil (dyed with Sudan B)
  o litmus cream
  o PHYSIOLOGY INTERACTIVE LAB SIMULATIONS
    (PH.I.L.S.) 4.0 and computer system (optional)

☐ **Carefully** read the directions and safety tips for this exercise before starting any procedure. **This set of exercises, in particular, requires careful preplanning of your time in the lab—or you may have difficulty finishing everything.**

A. Observing Carbohydrate Digestion

As you know, carbohydrates are molecules composed of carbon, hydrogen, and oxygen arranged in chemical units called **saccharides.** Polysaccharides are complex carbohydrates, such as *starch* and *glycogen,* composed of many saccharide units. **Disaccharides** such as *sucrose, lactose,* and *maltose,* are types of complex sugar composed of two saccharide units. **Monosaccharides,** such as the simple sugars *glucose (dextrose), galactose,* and *fructose,* are single saccharide units. Because the digestive tract absorbs carbohydrates only in their simplest form, monosaccharides, digestive processes must break polysaccharides and disaccharides into smaller units. Figure 34.1 presents a simplified scheme of how carbohydrates are digested in the human. **Amylases,** enzymes that break some of the bonds holding polysaccharides together, are found in the saliva and in pancreatic secretions. The disaccharides formed after the amylases have had their effect are then acted upon by **disaccharidase** enzymes to yield monosaccharides that can be absorbed into the blood.

In this exercise, you will use an extract of pancreatic secretions called *pancreatin*. Pancreatin contains pancreatic amylase, as well as pancreatic **lipase** and pancreatic **proteolytic** (protein-digesting) enzymes.

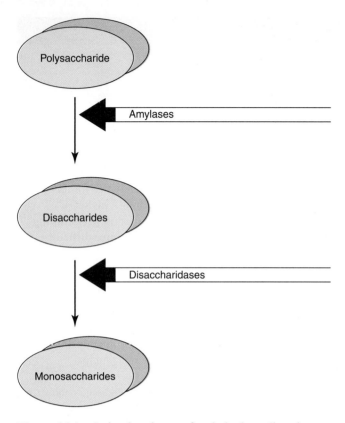

Figure 34.1 A simple scheme of carbohydrate digestion.

☐ **1** First, practice the tests used to analyze solutions for the presence of starch and its sugar products:
- **Lugol's test**—Lugol's test uses *Lugol's reagent* (an iodine solution) to test for the presence of starch. If Lugol's iodine turns from its normal orange color to a deep blue-black, starch is present. Mark a test tube **1a,** and put in 2 mL (two full droppers) of starch solution. Add a few drops of Lugol's reagent. Notice the color change. Anytime you see this result for Lugol's test, you have a positive result (meaning starch *is* present). Mark another tube **1b** and put in 2 mL distilled water and a few drops of Lugol's reagent. The solution should be a shade of orange, which is a negative result (meaning starch is *not* present). Save the tubes and their contents for later reference.

SAFETY FIRST!

Be careful of electrical and heat hazards when doing Benedict's test. Review safety procedures with the instructor before beginning this test.

- **Benedict's test**—Benedict's test uses Benedict's reagent to determine whether a solution contains certain types of sugar. Mark a test tube **2a** and add 2 mL of maltose solution. To that, add 2 mL of Benedict's reagent. In a tube marked **2b,** put 2 mL

distilled water and 2 mL Benedict's reagent. Place each tube in a beaker of boiling water on a hot plate for 3 minutes (fig. 34.2). At the end of 3 minutes, remove the tubes and place them in your rack. Tube **2a** should show a positive result, or a change from blue to another color, meaning sugar is present. Tube **2b** is an example of a negative result, with the contents remaining blue. Save these tubes for later reference.

> *Hint . . .* The chemical tests described here *will not work* if the test tubes are even slightly dirty. Make sure that your tubes are absolutely clean before each test.

☐ **2** Mark eight test tubes: **3a, 3b, 4a, 4b, 5a, 5b, 6a,** and **6b.** To each, add 2 mL starch. Next, have your lab partners help you simultaneously add a small amount of pancreatin (just enough powder to cover the end of the spatula) to each tube. Immediately
- Place tubes **3a** and **3b** in an ice water bath (0° to 1°C).
- Place tubes **4a** and **4b** in a rack at room temperature (20° to 25°C).
- Place tubes **5a** and **5b** in a warm water bath set at body temperature (37°C).
- Place tubes **6a** and **6b** in a boiling water bath (100°C).

Allow each tube to stand in its respective bath for 30 minutes. Remove tubes **6a** and **6b** from the boiling water earlier if they decrease to 0.5 mL or less.

☐ **3** At the end of 30 minutes, place all test tubes in the rack. Test all the **a** tubes for starch by adding a few drops of Lugol's reagent to each. Test all the **b** tubes for sugar by adding 2 mL Benedict's reagent to each and boiling them for 3 minutes. Record and interpret your results in Lab Report 34. Did starch digestion occur in any of the tubes? Which ones? How do you know? Did temperature affect digestion? How? Can you explain this result?

B. Observing Lipid Digestion

As described in figure 34.3, lipids undergo two phases of digestion. The first is a physical process called **emulsification.** Emulsification breaks up large fat droplets into small fat droplets, increasing the fat's surface area and allowing lipase molecules access to the lipid molecules. **Bile salts** secreted by the liver act as *emulsifying agents* in the human digestive tract. The second phase of lipid digestion is a chemical process involving *lipase*. Lipase catalyzes the breakdown of complex lipid molecules, such as **triglycerol** (triglyceride), into smaller units, such as **fatty acid** and **glycerol.** Pancreatin, an extract of pancreatic secretions, contains the digestive enzyme pancreatic lipase.

Figure 34.2 A boiling water bath, as used in Benedict's test for the presence of sugars.

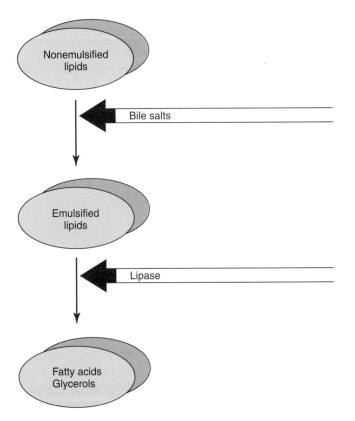

Figure 34.3 A summary of lipid digestion.

□ **1** Observe the emulsifying action of bile by performing this demonstration:
  o Mark a test tube **7a** and add 2 mL distilled water and 2 mL vegetable oil.

o Mark a test tube **7b** and add 2 mL distilled water, 2 mL vegetable oil, and 1 spatula of bile salts.
o Cap the tubes and shake them *simultaneously* for about 30 seconds.
o Compare the distribution of oil in both tubes. Continue watching them for about 10 minutes. How do you explain the difference between the two tubes?

□ **2** To observe the results of lipid digestion by pancreatic lipase, follow this procedure:
  o In a tube marked **8a,** put 2 mL litmus cream, 2 mL distilled water, and one spatula of pancreatin.
  o In a tube marked **8b,** put 2 mL litmus cream, 2 mL distilled water, and one spatula each of pancreatin and bile salts.
  o In a tube marked **8c,** put 2 mL litmus cream and 2 mL distilled water.
  o Place all three tubes in a 37°C water bath.
  o Litmus turns from blue to red in the presence of acids. If fatty acids form as a result of lipid digestion, the litmus cream (cream with litmus added) will turn from its original bluish tint to a lavender-pink color and eventually to pink. Observe your tubes for about an hour. Do any of the tubes turn pink? How long does it take? How do you explain these results? Report your findings in Lab Report 34.

Clinical Application

Pancreatin is an extract of pancreatic secretions. It includes pancreatic amylase, lipase, and proteolytic (protein-digesting) enzymes. Wearers of contact lenses sometimes use pancreatin solutions, marketed as "enzyme cleaners," to clean their lenses.

Contact lenses float on the tear fluid over the cornea. Over time, components of tear fluid build up on contact lenses. Tear fluid tends to form three layers, as shown in figure 34.4. The top layer is mostly lipids from skin oil (sebum) and other sources. The middle layer is an aqueous solution of salts, proteins, and other substances. The bottom layer is mainly composed of *mucin,* which in turn is composed of combined polysaccharide-protein molecules called *glycoproteins.*

With this background and what you already know about enzymes and digestion, answer the following application questions.

Figure 34.4   The three layers of tear fluid on the eye.

1. On which components of the tear fluid can pancreatin act? On what do you base your answer?

2. How would the digestion of tear fluid components help a wearer of contact lenses?

3. At what temperature would you recommend using enzymatic contact lens cleaning solutions? Explain.

4. Manufacturers of enzymatic contact lens cleaners recommend that the lenses be soaked in a nonenzyme solution for at least 4 hours after cleaning before wearing them again. What could happen if enzyme remained on a lens when put back on an eye?

LAB REPORT 34

Enzymes and Digestion

Name _____

Date _____ Section _____

Tube	Beginning Contents	Temp.	Test	Result	Explanation
1a	Starch	22°C	Lugol's (starch)		
1b	Water	22°C	Lugol's (starch)		
2a	Maltose	22°C	Benedict's (sugar)		
2b	Water	22°C	Benedict's (sugar)		
3a	Starch, pancreatin	0°C	Lugol's (starch)		
3b	Starch, pancreatin	0°C	Benedict's (sugar)		
4a	Starch, pancreatin	22°C	Lugol's (starch)		
4b	Starch, pancreatin	22°C	Benedict's (sugar)		
5a	Starch, pancreatin	37°C	Lugol's (starch)		
5b	Starch, pancreatin	37°C	Benedict's (sugar)		
6a	Starch, pancreatin	100°C	Lugol's (starch)		
6b	Starch, pancreatin	100°C	Benedict's (sugar)		
7a	Water, oil	22°C	Visual observation		
7b	Water, oil, bile salts	22°C	Visual observation		
8a	Litmus cream, water, pancreatin	37°C	Litmus (acid)		
8b	Litmus cream, water, pancreatin, bile salts	37°C	Litmus (acid)		
8c	Litmus cream, water	37°C	Litmus (acid)		

Discuss the overall meaning of your results by answering the questions posed in step 3 of activity A:

Discuss the overall meaning of your results by answering the questions posed in steps 1 and 2 of activity B:

Fill-in

_____ 1

_____ 2

_____ 3

_____ 4

_____ 5

_____ 6

_____ 7

_____ 8

_____ 9

_____ 10

_____ 11

_____ 12

_____ 13

_____ 14

_____ 15

Fill-in

1. Lipase is an enzyme that digests __?__ molecules.
2. Amylase is an enzyme that digests __?__ molecules.
3. Bile is helpful in preparing __?__ for digestion.
4. The __?__ secretes bile salts.
5. Polysaccharides are broken apart to yield __?__ molecules before being completely digested.
6. Carbohydrates are absorbed into the body in the form of __?__ molecules.
7. The substance __?__ contains several enzymes and is often used to clean contact lenses.
8. __?__ test is used to detect the presence of starch.
9. __?__ test is used to detect the presence of certain sugars.
10. __?__, here used in fatty cream, is a substance used to detect the presence of acids.
11. Enzymes that break apart proteins are termed __?__ enzymes.
12. Glucose is an example of a(n) __?__ type of carbohydrate molecule.
13. Bile __?__ globules of fat.
14. A(n) __?__ is a protein that catalyzes biochemical reactions.
15. Enzymes called __?__ digest sugars, such as sucrose and maltose.

Advanced Scientific Inquiry
PHYSIOLOGY INTERACTIVE LAB SIMULATION (PH.I.L.S. 4.0)

If available, use the computerized laboratory simulation PH.I.L.S. 4.0 to simulate one or more of the following activities that explore digestive function:

- *Glucose Transport (simulation #41)*

For each of the activities you perform, print the lab report according to the directions given in the activity and attach a copy of all sheets to your lab report (or as your instructor directs).

Urinary Structures

The **urinary system** is composed of the kidneys, ureters, urinary bladder, urethra, and associated structures. The functions of the urinary system include maintaining extracellular fluid balance, excreting wastes, and maintaining blood pH.

In this exercise, you are challenged to discover the basic anatomy of the urinary system on both the gross and microscopic levels. In Lab Exercise 36, you will analyze the components of the **urine** produced by this system.

Before You Begin

☐ Read the appropriate chapter in your textbook.

☐ Set your learning goals. When you finish this exercise, you should be able to:
- identify the major organs of the urinary system and find them in models and charts
- describe the gross anatomical features of the kidney and identify them in figures, models, and specimens
- describe the features of the nephron and locate them in figures and models
- identify the renal corpuscle in a prepared microscopic specimen

☐ Prepare your materials:
- model of the human torso (dissectible)
- model of the kidney (frontal section)
- preserved sheep kidney (double or triple injected, if available)
- dissection tools and trays
- model of a nephron and associated structures
- microscope
- prepared microslide: *kidney cortex c.s.*

☐ **Carefully** read the directions and safety tips for this exercise before starting any procedure.

A. The Urinary Plan

Study the layout of the urinary system by locating these features in a dissectible model of the human torso (fig. 35.1):

☐ **1** The left and right **kidneys** are located behind the peritoneum, along the posterior abdominal wall. Locate these kidney features:
- **renal fat pad**—(not shown in all models) tissue that surrounds and protects each kidney
- **renal capsule**—the fibrous outer wall of the kidney
- **hilum**—an indentation on the medial side of each kidney where vessels and nerves enter or exit

☐ **2** The **ureter** is a muscular tube that exits each kidney at the hilum and extends posteriorly to the pelvic cavity.

☐ **3** The **urinary bladder** is a collapsible, muscular sac for the temporary storage of urine. The two ureters enter on each side of the posterior floor of the bladder.

☐ **4** The **urethra** is a muscular tube that extends from the anterior floor of the bladder to the outside of the body. In the female, it is a short tube that ends just anterior to the vagina. In the male, it is much longer, extending all the way through the penis. In the male, it conducts semen as well as urine.

Urine is formed in each kidney and is conducted through the ureters to the bladder. When the bladder is full and it is convenient, **urinary sphincters** that control flow through the urethra relax and allow urine to exit the body.

B. The Kidney Model

Locate these features of the human kidney on a model (or chart) of a frontal section:

The Urinary System

KIDNEY 1
URETER 2
URINARY BLADDER 3
URETHRA 4
RENAL CAPSULE 5
HILUM 6
RENAL SINUS 7
RENAL PELVIS 8
CALYCES 9
RENAL CORTEX 10
RENAL MEDULLA 11
RENAL PYRAMIDS 12
RENAL ARTERY 13
RENAL VEIN 14
NEPHRON 15

Figure 35.1 Use colored pens or pencils to color the labels and the features of the urinary system.

☐ **1** Locate the **hilum** and **renal capsule** from this perspective.

☐ **2** Identify these structures located just within the hilum:
- ○ **renal sinus**—a fat-filled cavity
- ○ **renal pelvis**—a wide section of the urinary channel, distal to the ureter
- ○ **calyces**—branches of the pelvis extending from the kidney tissue proper

☐ **3** The kidney tissue proper is divided into an outer **cortex** and an inner **medulla**, as is the tissue of many organs. **Renal pyramids** are cone-shaped sections of tissue lying mostly within the medulla. Each appears as a triangle in a frontal section. Each is composed of collecting ducts that conduct urine toward its tip (papilla), which is surrounded by the end of a calyx.

Urine formed by the kidney is conducted through the collecting ducts of the renal pyramids to the calyces. The calyces fuse to form the large renal pyramid. As urine is conducted out of the kidney through the hilum, the urinary channel narrows to form the ureter.

☐ **4** Identify these features of the renal arterial supply (fig. 35.2):

- ○ **renal artery**—a branch of the abdominal aorta entering the hilum and extends through the renal sinus
- ○ **interlobar arteries**—branches of the renal arteries that extend outward through the tissue between the pyramids
- ○ **arcuate arteries**—branches of the interlobar arteries that turn to extend between the cortex and medulla
- ○ **interlobular arteries**—branches of the arcuate arteries that extend outward, into the cortex
- ○ **afferent arterioles**—small arteries that arise from branches of interlobular arteries, each one extending to a glomerulus
- ○ **glomerular capillaries**—capillaries arising from an afferent arteriole and forming a ball or small capillary bed called a **glomerulus**

☐ **5** After a brief detour, the venous network of the kidney parallels the arterial network, as you may expect:
- ○ **efferent arterioles**—each extending from a glomerulus
- ○ **peritubular capillaries**—small capillary beds arising from an efferent arteriole (a network of peritubular capillaries surrounds each

Figure 35.2 Identify the renal blood vessels indicated by label lines.

nephron, the tubular, microscopic unit of the kidney)
- ○ **interlobular veins**
- ○ **arcuate veins**
- ○ **interlobar veins**
- ○ **renal veins**   blood vessels extending through the renal sinus and out the hilum, draining into the inferior vena cava

Hint . . . Double-injected specimens have red latex injected into the arteries and blue latex injected into the veins. Triple-injected specimens also have yellow latex injected into the urinary channels and tubules. You may be able to see some of the finer detail visible in an injected specimen by using a hand lens or a dissection microscope.

C. The Sheep Kidney

SAFETY FIRST!

Observe the usual precautions when dissecting preserved specimens. Be sure to follow the safety advice that accompanies the preservative used in your specimen. Avoid injury with the dissection tools.

The sheep kidney is very similar to the human kidney and makes an ideal specimen for study.

☐ **1**   Examine the external aspect of your specimen. Identify as many parts of the kidney as you can. Refer to the aid given in activities A and B if you have trouble. You may have to remove some of the renal fat pad.

☐ **2**   Use a long knife or scalpel to cut a section, dividing the kidney into roughly equal dorsal and ventral portions (as in figure 35.3). Try to identify as many features as you can.

D. The Nephron Model

The functional units of the kidney are the tiny tubules called **nephrons.** The nephrons carry out the three basic processes that accomplish the kidney's function, forming *urine* as they do so: **filtration, tubular reabsorption,** and **tubular secretion.** Material from the blood is filtered into the beginning of the nephron. As the *filtrate* moves along the nephron tubule, some substances are reabsorbed into the blood, and some additional substances are secreted from the blood into the filtrate. The urine thus produced is channeled to collecting ducts, which drain it from the kidney. Locate the main features of the nephron and associated structures in a model or chart (fig. 35.4).

☐ **1**   The **renal corpuscle** is the roughly spherical structure at the beginning of the nephron. The inner portion of the renal corpuscle is the **glomerulus,** a ball of glomerular capillaries. Surrounding the glomerulus is the double-walled **Bowman's capsule.**

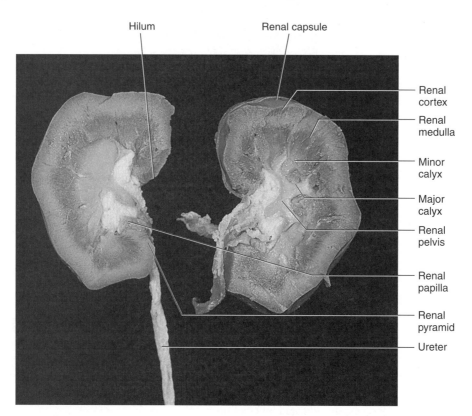

Figure 35.3   Dissected sheep kidney.

Identify the afferent and efferent arterioles in the model. Can you trace their path into the renal corpuscle, out, then to the peritubular capillaries?

□ **2** The **proximal convoluted tubule** is a narrow channel proceeding from the Bowman's capsule of the renal corpuscle. It is *convoluted,* meaning that it is coiled.

□ **3** Filtrate formed in the Bowman's capsule flows through the proximal convoluted tubule and into the **loop of Henle.** Anatomically, the loop is a continuation of the proximal tubule. In many nephrons, the **descending limb** of the loop dips far down into the medulla, turns, and returns as the **ascending limb** to the cortex.

□ **4** Filtrate flows from the ascending limb of the loop of Henle into the **distal convoluted tubule.** The filtrate is emptied into a **collecting duct,** a tubule that collects urine from many nephrons and conducts it through a renal pyramid to a calyx. Eventually, the urine is voided during **urination** or **micturition.**

□ **5** Your model may represent a nephron whose distal convoluted tubule passes between the afferent arteriole and efferent arteriole, near the renal corpuscle. If so, you may notice that the afferent arteriole wall and distal tubule wall form a specialized structure where they meet. This structure is called the **juxtaglomerular apparatus.** (*Juxtaglomerular* means "near the glomerulus.")

Figure 35.4 Label the indicated parts of the nephron.

Nephron tubules

Renal corpuscle
Glomerulus
Bowman's capsule

Figure 35.5   Micrograph of a renal cortex cross section. A renal corpuscle and surrounding tubules are visible.

The juxtaglomerular apparatus secretes **renin,** an enzyme that catalyzes the conversion of *angiotensinogen* to *angiotensin I.* This begins a series of steps that help regulate blood pressure.

E. Microscopic Specimen

Obtain a prepared microscopic specimen of a cross section of renal cortical tissue (fig. 35.5). Scan it under low power first and locate one or more renal corpuscles, which are easily identifiable dark circles. Switch to high power and try to identify the features listed.

SAFETY FIRST!

Observe the usual precautions when using the microscope and a prepared slide.

☐ **1**   The **glomerulus** is the dark region forming the center of the renal corpuscle. It is a network of glomerular capillaries.

☐ **2**   The **Bowman's capsule** is seen as a very thin white or light area surrounding the glomerulus. Actually,

the *lumen* of the capsule appears white. The parietal wall of the Bowman's capsule is composed of simple squamous epithelium, which can sometimes be distinguished. The visceral wall of the Bowman's capsule is composed of specialized epithelial cells called **podocytes.** Podocytes have extensions that wrap around the capillary walls to form a **filtration membrane.** You will not be able to distinguish podocytes in your specimen.

☐ **3**   The tissue surrounding each renal corpuscle is mostly composed of **renal tubules.** You will not be able to distinguish which are slices of proximal tubules or distal tubules, but you can appreciate the single-cell structure of their walls. What type of epithelium forms the tubule walls? How is this type of epithelial tissue specialized to perform the functions of the nephron tubule?

Hint . . . If you used double- or triple-injected sheep kidney specimens in the dissection, you may be able to see renal corpuscles and nephron tubules by examining a sectioned specimen with a hand lens or dissection microscope.

LAB REPORT 35

Name _____

Urinary Structures

Date _____ Section _____

Structure	Sheep	Human	Function(s)
Kidney	☐		
Renal capsule	☐	☐	
Hilum	☐	☐	
Ureter	☐	☐	
Urinary bladder		☐	
Urethra		☐	
Renal sinus	☐	☐	
Renal pelvis	☐	☐	
Calyces	☐	☐	
Renal cortex	☐	☐	
Renal medulla	☐	☐	
Renal pyramids	☐	☐	
Renal artery, vein	☐	☐	
Interlobar arteries, veins	☐	☐	
Arcuate arteries, veins	☐	☐	
Interlobular arteries, veins	☐	☐	
Afferent arteriole	☐	☐	
Glomerulus	☐	☐	
Efferent arteriole	☐	☐	
Peritubular capillaries	☐	☐	
Renal corpuscle	☐	☐	
Bowman's capsule	☐	☐	
Proximal convoluted tubules	☐	☐	
Loop of Henle: descending limb, ascending limb	☐	☐	
Distal convoluted tubule	☐	☐	
Collecting duct	☐	☐	
Juxtaglomerular apparatus		☐	

Specimen: *renal cortex c.s.*
        Total magnification: _____

Specimen: *renal cortex c.s.*
        Total magnification: _____

Figure 35.2 *(from p. 259)*

_____ 1

_____ 2

_____ 3

_____ 4

_____ 5

_____ 6

_____ 7

_____ 8

_____ 9

Figure 35.4 *(from p. 261)*

_____ 1

_____ 2

_____ 3

_____ 4

_____ 5

_____ 6

_____ 7

_____ 8

_____ 9

_____ 10

_____ 11

_____ 12

Put in order

_____ 1

_____ 2

_____ 3

_____ 4

_____ 5

_____ 6

_____ 7

_____ 8

_____ 9

_____ 10

_____ 11

Fill-in

_____ 1

_____ 2

_____ 3

_____ 4

_____ 5

_____ 6

_____ 7

_____ 8

_____ 9

_____ 10

Put in order

Arrange these structures in the order in which substances flow through the urinary system, beginning with the site of filtration.

ascending limb of the loop of Henle

Bowman's capsule

calyces

collecting duct

descending limb of the loop of Henle

distal convoluted tubule

proximal convoluted tubule

renal pelvis

ureter

urethra

urinary bladder

Fill-in

1. The ___?___ is a fat-filled cavity just inside the kidney's hilum.
2. The renal ___?___ are branches of the renal pelvis.
3. The ___?___ is a muscular tube extending from the renal pelvis to the urinary bladder.
4. The three basic processes observed in the nephron are filtration, tubular reabsorption, and tubular ___?___.
5. The ___?___ conducts blood into the glomerular capillaries.
6. The ___?___ conducts blood into the kidney.
7. The ___?___ capillaries surround the tubules of the nephron.
8. The inner tissue of the kidney is termed the renal ___?___.
9. The ___?___ apparatus secretes the enzyme renin.
10. The collecting ducts converge as they extend toward the calyces, forming the renal ___?___.

LAB EXERCISE 36

Urinalysis

Urinalysis is the examination of urine and its contents. Urinalysis can be very comprehensive, testing for dozens of different physical and chemical characteristics. On the other hand, some clinical situations call for urinalysis that tests only one or two urine characteristics. For example, the pregnancy test and ovulation test performed in Lab Exercise 24 were types of urinalysis that each determined the presence or concentration of one particular hormone.

In this exercise, you will perform some of the more common clinical urinalysis tests, using the most current methods.

Before You Begin

☐ Read the appropriate chapter in your textbook.

☐ Set your learning goals. When you finish this exercise, you should be able to:
  ○ state the normal characteristics of freshly voided urine
  ○ demonstrate the use of dip-and-read clinical test strips for urinalysis
  ○ explain the significance of common urine tests
  ○ prepare and examine a stained urine sediment slide
  ○ evaluate urinalysis results

☐ Prepare your materials:
  ○ fresh urine specimen (your own or a packaged substitute) in a disposable container
  ○ dip-and-read multiple test strips (Multistix 10SG are preferred)
  ○ urine centrifuge and tapered tubes
  ○ disposable 1 mL droppers
  ○ Sedistain urine sediment stain
  ○ paper towels and wipes
  ○ microscope slides and coverslips
  ○ microscope
  ○ full-color urine sediment chart (optional)
  ○ BIOHAZARD container
  ○ unknown urine additives (optional)
  ○ PHYSIOLOGY INTERACTIVE LAB SIMULATIONS (PH.I.L.S.) 4.0 and computer system (optional)

☐ **Carefully** read the directions and safety tips for this exercise before starting any procedure.

Examination of Urine

Urine reflects the overall status of extracellular fluid because it is derived from blood, and its contents have been adjusted to some extent on the basis of homeostatic balance. Additionally, the health of the kidney and urinary tract in particular affects the characteristics of urine. Observe the urinalysis tests described, or perform them yourself on *your own urine* or a packaged substitute provided by the instructor.

SAFETY FIRST!

Assume that all body fluids contain disease-causing agents. Protect yourself and others from contamination by wearing protective lab apparel, gloves, and eyewear. Disinfect all surfaces that have, or *could have,* come into contact with urine. *Immediately* put all disposable urine containers, droppers, towels, wipes, slides, and coverslips in a BIOHAZARD container after use. Your instructor may prefer to do this exercise as a demonstration.

☐ **1** Examine these physical characteristics of urine after placing some of the sample in a transparent container:
  ○ **Transparency**—Normal urine is clear to slightly cloudy (especially after standing). Cloudy urine may contain fat globules, epithelial cells, mucus, microbes, or chemicals.
  ○ **Color**—Normal urine is amber, straw, or transparent yellow due to the presence of **urochromes,** such as *urobilinogen* (a bacterial product derived from *bilirubin*). Yellow-brown to greenish urine may occur when a high concentration of bile pigments are present. A red-to-dark-brown color may indicate the presence of blood. These or other abnormal colors may also result from the presence of food pigments (carotene, for example), drugs, or other chemicals.

☐ **2** Many characteristics of urine are determined by the use of paper that is impregnated with test reagents. The paper is dipped into a sample, and a color change indicates the presence (and sometimes concentration) of a particular substance. You may have used pH test paper in your aquarium or in another lab course. In this step, you will use a plastic strip on which 9 or 10 test papers have been placed. The papers are impregnated with different reagents.

> *Hint...* Carefully read the instructions provided by the distributor of your test strips. Follow the instructions *exactly,* especially those dealing with the times at which test strips are to be read.

Shake or stir the sample so that any sediment on the bottom of the container becomes suspended. With gloved hands, dip the papered end of one test strip into your sample. Lift it out and tap the excess urine onto the inside rim of the container. You may need to transfer some urine to a test tube so that you can immerse all the paper pads. Read the results by looking at the color of each pad in order (fig. 36.1), comparing the test papers with standard charts as instructed. Usually, certain tests *must* be read at a certain point in time. A 10-test strip tests for these urine components:

- ○ **Leukocytes**—Occasional WBCs are normal, but values increase in urinary infections.
- ○ **Nitrite**—A positive nitrite result indicates the presence of large amounts of bacteria, as in an infection. This test is useful if one has a cloudy urine sample but no microscope to determine whether bacteria are causing the cloudiness.
- ○ **Urobilinogen**—A derivative of bilirubin, high levels of urobilinogen may indicate excessive RBC destruction or liver disease.
- ○ **Protein**—*Albumin* is a small protein molecule that is normally absent or present only in trace amounts. Higher levels may indicate hypertension or kidney disease. Detectible levels are normal just after exercise.

- ○ **pH**—The relative H^+ concentration, pH is a determinant of acidity. Normal urine pH is 4.6 to 8. Values are lower (more acid) in acidosis, starvation, and dehydration. pH is higher (more alkaline) in urinary infections and alkalosis.
- ○ **Occult blood** (RBCs or free hemoglobin)—*Occult* means hidden. This strip tests for small amounts of hemoglobin that don't discolor the urine but are still clinically significant. Normally not present, hemoglobin may indicate kidney infection or the presence of stones in the kidney, ureter, or bladder.
- ○ **Specific gravity**—This is the ratio of urine density to water density. If urine is pure water, the specific gravity is 1.000. Normal urine is 1.001 to 1.030. Lower values may indicate kidney disease. Higher values indicate high solute concentration and may occur during dehydration or diabetes mellitus. (This test is not present in 9-test strips.) An alternate method for determining specific gravity is to float a **hydrometer** in a cylinder containing urine, as in figure 36.2. Read the scale on the hydrometer at the point that it meets the surface of the urine.

Figure 36.1   Urine test strip. Read the strip near the end, where you hold onto the strip with your fingers, and progress to the other end of the strip.

Figure 36.2   The specific gravity of urine can be determined by floating a hydrometer in room-temperature urine.

o **Ketone**—A by-product of fat metabolism, it may be present during fasting, diabetes mellitus, or a low-carbohydrate diet.

o **Bilirubin**—Normally not present, or present in trace amounts, this product of RBC destruction in the liver may indicate liver disease or bile tract obstruction if present in the urine.

o **Glucose**—Normally, there is no glucose in the urine. Trace amounts may be present after eating a meal high in carbohydrates. Continued high levels in the urine may indicate diabetes mellitus or pituitary problems.

SAFETY FIRST!

Don't forget to observe the usual precautions when using the microscope and stained, wet-mount slides.

☐ **3** Observe the components of urine sediment by performing this procedure:

o Shake or stir the urine sample to suspend any sediment that has settled.

o Transfer some of the urine to a tapered centrifuge tube.

o After balancing and securing the centrifuge as your instructor demonstrates, spin the sample for about 5 minutes.

o Without disturbing the sediment that has collected in the tip of the tube, squeeze the bulb of a disposable dropper and gently lower it into the urine. When the dropper reaches the bottom, release the bulb *slightly* to collect only a few drops of sediment-containing urine.

o Quickly place a drop of the sample on a clean microscope slide. Add one drop of Sedistain and cover it with a coverslip. Remove excess fluid from the edges with a lab wipe and immediately discard the wipe in a BIOHAZARD container.

o Examine the specimen under both low and high power. Move from field to field as you identify different types of sediment.

☐ **4** Your instructor may have a chart of sediment types for you to use in identifying urine solids. Figures 36.3 and 36.4 show examples of common urine sediment components:

o **Cells**—Epithelial cells from the urinary tract lining, blood cells from injury or infection sites, or infectious microbes may be present in sediment. If more than trace amounts of blood or microbial cells are present, a urinary problem is indicated.

o **Artifacts**—Artifacts are materials that have accidentally gotten into the sample, including fabric fibers from underwear, powder used on the skin near the urethral opening, or skin oil droplets.

o **Crystals**—Very tiny crystals of normal urine components or drugs may be visible under high power. Small amounts of these crystals may be normal. A large number of crystals is seen in urinary *retention*, the inability to void urine from the bladder. Large masses of crystals are called *stones,* or *calculi.*

o **Casts**—Casts are chunks of material that have hardened somewhere in the urinary channel and sloughed off into the urine. They may be roughly cylindrical masses of cells, granules, or other substances.

☐ **5** Your instructor may opt to offer unknown urine additives. Add the additive to your sample as instructed and test the sample as if it were a new urine specimen. The additive will have altered the sample's characteristics in a way that mimics an abnormal condition. Your instructor may choose instead to give you a new sample, one that is abnormal in some respect. Test the new sample in the same way you tested the first sample. Report your results in Lab Report 36.

Advanced Scientific Inquiry
PHYSIOLOGY INTERACTIVE LAB SIMULATION (PH.I.L.S. 4.0)

If available, use the computerized laboratory simulation PH.I.L.S. 4.0 to simulate the effects of antidiuretic hormone (ADH) on urine production:

• *Anti-Diuretic Hormone (simulation #42)*

Print the lab report according to the directions given in the activity and attach a copy of all sheets to your lab report (or as your instructor directs).

Cells

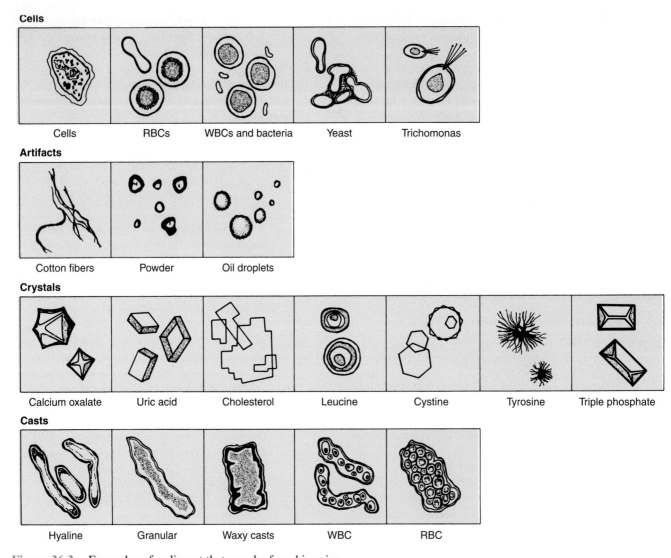

Figure 36.3   Examples of sediment that may be found in urine.

Figure 36.4   Micrographs of examples of different urine sediment types.

LAB REPORT 36

Urinalysis

Name _____

Date _____ Section _____

Characteristics	Normal Value	Observed Value	Interpretation
Transparency			
Color			
Leukocytes			
Nitrite			
Urobilinogen			
Protein			
pH			
Occult blood			
Specific gravity			
Ketone			
Bilirubin			
Glucose			

Summary and evaluation of this sample:

Sketch (and label) examples of urine sediment found in your sample:

Use this table for results of the test on the "unknown" sample:

UNKNOWN CODE _____

Characteristic	Normal Value	Observed Value	Interpretation
Transparency			
Color			
Leukocytes			
Nitrites			
Urobilinogen			
Protein			
pH			
Occult blood			
Specific gravity			
Ketone			
Bilirubin			
Glucose			

Summary and evaluation of this sample:

LAB EXERCISE 37

The Male Reproductive System

Humans, as a species, have two **reproductive systems:** a male system and a female system. The primary function of both is the sexual reproduction of offspring. Sexual reproduction in humans requires the successful fusion of two **gametes,** or sex cells. Human gametes, **sperm** and **ova,** are formed by means of *meiosis* and so each has 23 chromosomes. When a sperm cell from the male parent unites with an ovum from the female parent, a cell with 46 chromosomes is formed. This offspring cell has a unique mix of DNA from both parents, providing the variation essential to the survival of the human species.

In this exercise, you are invited to explore the structure and function of the male reproductive system. In Lab Exercise 38, the female reproductive system is presented.

Before You Begin

☐ Read the appropriate chapter in your textbook.

☐ Set your learning goals. When you finish this exercise, you should be able to:
  ○ describe the plan of the male reproductive tract
  ○ describe the organs of the male reproductive system and their principal functions
  ○ locate male reproductive organs in charts and models
  ○ identify major features in a microscopic specimen of the testis
  ○ identify the features of mature sperm in figures and in a prepared sperm smear

☐ Prepare your materials:
  ○ models and charts of the male reproductive system
  ○ microscope
  ○ prepared microslides: *human testis c.s.*
                          human sperm smear

☐ **Carefully** read the directions and safety tips for this exercise before starting any procedure.

A. Human Model

Find the major features of the male reproductive system in models and charts (fig. 37.1).

☐ **1** The primary sex organs of the male reproductive system are the two **testes.** The testes are located within the **scrotum,** a sac of skin and other tissues on the outer wall of the anterior trunk. A thin layer of smooth muscle called the **dartos muscle** forms part of the scrotum's wall. Find these structures associated with the testes:
  ○ **Capsule**—A connective tissue outer wall, the capsule has inward extensions that divide each testis into **lobules.**
  ○ **Seminiferous tubules**—They are long, coiled tubules in which sperm cells are produced. Tissue between the tubules contains **interstitial cells (of Leydig),** which secrete *testosterone.*
  ○ **Rete testis**—It is a network of tubules into which the seminiferous tubules empty. The rete testis, in turn, empties into 15 to 20 **efferent ductules** that leave the testis.

☐ **2** The **epididymis** is a set of coiled tubules on the outside of the testis (but still within the scrotum). Sperm cells formed in the testis move into the epididymis, where they continue to mature.

☐ **3** The **ductus deferens,** or **vas deferens,** conducts sperm cells from the epididymis out of the scrotum and into the pelvic cavity. Trace its path in your model. As the ductus deferens enters the abdominopelvic cavity through the *inguinal canal,* it is associated with nerves, vessels, and a connective tissue covering. Together, these structures form the **spermatic cord.** Near its end, each ductus deferens has an enlarged section called the **ampulla.**

☐ **4** Joining the ductus deferens at the ampulla is a duct from a **seminal vesicle.** Each seminal vesicle is a gland that produces fluid that becomes part of **semen.** Semen is the fluid medium containing sperm that is *ejaculated* during the male sexual response. The ductus deferens and seminal vesicle duct unite to form the **ejaculatory duct,** which extends to the urethra, just inferior to the bladder.

☐ **5** Surrounding the urethra and ejaculatory ducts below the bladder is the **prostate gland.** This gland, like the seminal vesicles, contributes to the seminal fluid.

☐ **6** Sperm-containing semen is conducted through the **urethra** during the male sexual response. Near the base of the penis, ducts from the **bulbourethral**

The Male Reproductive System

SCROTUM 1
TESTIS 2
EPIDIDYMIS 3
DUCTUS DEFERENS 4
SEMINAL VESICLE 5
EJACULATORY DUCT 6
PROSTATE GLAND 7
URETHRA 8
BULBOURETHRAL GLAND & DUCT 9
CORPUS SPONGIOSUM 10
  GLANS PENIS 11
CORPORA CAVERNOSA 12
FORESKIN 13
URINARY BLADDER 14
PUBIS 15
RECTUM 16

Figure 37.1   Color the labels and the features of the male reproductive system with colored pens or pencils.

glands join the urethra. These small glands contribute a small amount of fluid to the semen.

☐ 7 The **penis** is the erectile structure through which the urethra conducts semen out of the body during ejaculation. Erection of the penis is essential to insertion into the female tract and deposition of the semen there. Three vascular bodies, the left and right **corpora cavernosa** and the **corpus spongiosum,** engorge with blood during the sexual response and stiffen the penis. The tip of the corpus spongiosum forms the head of the penis, or **glans penis.** The glans penis is covered by a fold of skin called the **foreskin,** or **prepuce,** unless it has been removed by means of *circumcision.*

SAFETY FIRST!

The remaining activities of this exercise call for the use of a microscope. Be sure to observe the usual precautions.

B. Microscopic Structure of the Testis

First in a model or chart, then in a prepared microscopic specimen, locate these structures in a cross section of the testis (fig. 37.2):

☐ 1 The major feature of a testis cross section is the presence of numerous **seminiferous tubules.** Identify the **basement membrane,** lumen, and walls of a seminiferous tubule. Locate some of the **interstitial cells** between the tubules.

☐ 2 Examine the cells forming the wall of a seminiferous tubule. Large, pale cells with oval nuclei are supportive **nurse (Sertoli) cells,** or **sustenacular cells.** They extend all the way to the lumen, but their entire length is often difficult to distinguish. The majority of cells are **germ cells.** The outermost layer of germ cells are **spermatogonia.** Some daughter cells of the spermatogonia, nearer the lumen, become **spermatocytes** that divide by means of meiosis to form **spermatids.** Each spermatid develops to become a **spermatozoan,** many of which are seen in the lumen.

C. Sperm Cells

Examine a prepared microscopic specimen of a sperm (semen) smear under high power (fig. 37.3). Notice that each normal sperm has a pear-shaped **head** containing a nucleus. A **midpiece** containing mitochondria is at the base of a **tail,** or *flagellum* (fig. 37.4).

Up to one-third of the sperm cells in a normal specimen may be deformed. Look for sperm with multiple or misshapen heads or with tail deformities. High ratios of deformed sperm are associated with stress, infection, or high environmental temperatures.

Figure 37.2 Micrograph of a testis cross section.

(a)

(b)

Figure 37.3   Human sperm smear (with Papanicolaou Stain). (*a*) Low power (200x), (*b*) High power (400x).

1

2

3

4

5

6

7

8

9

10

11

(a)

12

13

14

15

16

17

(b)

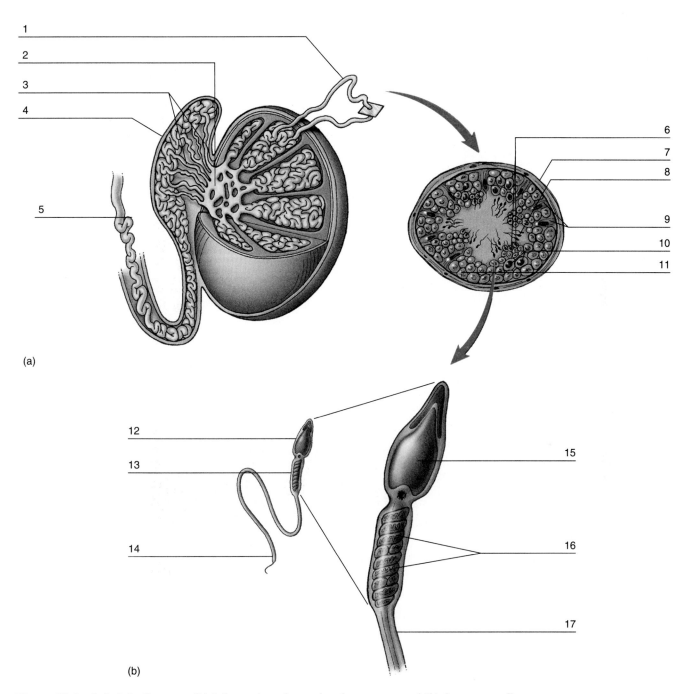

Figure 37.4 Label the features of (*a*) the testis and associated structures and (*b*) the sperm cell.

Spermatogenesis

Spermatogenesis is the process by which spermatozoa develop within the wall of the seminiferous tubule of the testis. Figure 37.5*a* shows a flow chart of the main events of meiotic division that produce sperm cells. Notice that the number of chromosomes of the original parent cell (spermatogonium, or *germ cell*) is reduced from 46 to 23 in the daughter sperm cells. Figure 37.5*b* shows the relative position of each cell stage within the wall of the seminiferous tubule inside the testis.

1. Spermatogonia are the cells from which sperm cells arise. The spermatogonia divide by mitosis. One daughter cell remains a spermatogonium that can divide again by mitosis. One daughter cell becomes a primary spermatocyte.

2. The primary spermatocyte divides by meiosis to form secondary spermatocytes.

3. The secondary spermatocytes divide by meiosis to form spermatids.

4. The spermatids differentiate to form sperm cells.

Figure 37.5   Spermatogenesis. (*a*) Meiosis diagrammed. (*b*) Cross section of a seminiferous tubule, showing the process of spermatogenesis.

The Male Reproductive System

Name _____

Date _____ Section _____

Structure	Model or Chart	Function(s)
Testis	☐	
Scrotum	☐	
Capsule	☐	
Seminiferous tubule	☐	
Rete testis	☐	
Epididymis	☐	
Ductus deferens	☐	
Seminal vesicle	☐	
Ejaculatory duct	☐	
Prostate gland	☐	
Urethra	☐	
Bulbourethral gland	☐	
Penis	☐	
Corpora cavernosa	☐	
Corpus spongiosum	☐	
Glans penis	☐	
Foreskin	☐	
Seminiferous tubule	☐	
Interstitial cell	☐	
Sertoli cell	☐	
Germ cell	☐	
Spermatogonium	☐	
Spermatocyte	☐	
Spermatid	☐	
Spermatozoan: head, midpiece, tail	☐	

Specimen: *human testis c.s.*
Total magnification: _____

Specimen: *human sperm smear*
Total magnification: _____

Figure 37.3 *(from p. 274)*

_____ 1

_____ 2

_____ 3

_____ 4

_____ 5

_____ 6

_____ 7

_____ 8

_____ 9

_____ 10

_____ 11

_____ 12

_____ 13

_____ 14

_____ 15

_____ 16

_____ 17

Multiple choice

_____ 1

_____ 2

_____ 3

_____ 4

_____ 5

_____ 6

Multiple choice

1. Semen travels through the male reproductive tract in this order:
   a. ejaculatory duct, ductus deferens, epididymis, urethra
   b. epididymis, ductus deferens, ejaculatory duct, urethra
   c. urethra, ejaculatory duct, ductus deferens, cpididymis
   d. ductus deferens, epididymis, ejaculatory duct, urethra

2. Sperm cells mature in phases that occur in this order:
   a. spermatogonium, spermatocyte, spermatozoan, spermatid
   b. spermatocyte, spermatozoan, spermatid, spermatogonium
   c. spermatogonium, spermatocyte, spermatid, spermatozoan

3. Erection of the penis is accomplished through increased blood volume in
   a. the corpus spongiosum.
   b. the foreskin.
   c. the corpora cavernosa.
   d. *a* and *c*.
   e. none of the above.

4. Mitochondria of the sperm cell are found in the
   a. head.
   b. midpiece.
   c. tail.

5. The urethra of the male functions as a part of
   a. the urinary system.
   b. the digestive system.
   c. the reproductive system.
   d. *a* and *c*.
   e. all of the above.

6. The distal, widened portion of the ductus deferens is more specifically known as the
   a. ampulla.
   b. vas deferens.
   c. seminal vesicle.
   d. prostate gland.
   e. proximal tubule.

The Female Reproductive System

This exercise continues the study of human reproductive systems begun in Lab Exercise 37. In this exercise, you are invited to explore the essential structure and function of the female reproductive system.

Before You Begin

☐ Read the appropriate chapter in your textbook.

☐ Set your learning goals. When you finish this exercise, you should be able to:
  ○ describe the plan of the female reproductive tract
  ○ describe the organs of the female reproductive systems and their principal functions
  ○ locate female reproductive organs in charts and models
  ○ identify major features in a microscopic specimen of a mammalian ovary
  ○ identify the structures of the mammary glands

☐ Prepare your materials:
  ○ models and charts of the female reproductive system and the breasts
  ○ microscope
  ○ prepared microslide: *mammalian ovary c.s.*

☐ **Carefully** read the directions and safety tips for this exercise before starting any procedure.

A. Human Model

Find the major features of the female reproductive system in models and charts.

☐ **1**   The primary sex organs of the female reproductive system are the two **ovaries.** They are small, rounded organs suspended in the pelvic cavity by ligaments. The ovarian mesentery, or **mesovarium,** attaches the ovaries to the **broad ligament.** The **suspensory ligaments** and **ovarian ligaments** also support the ovaries.

☐ **2**   A **uterine tube** is associated with each ovary. Also called **fallopian tube,** or **oviduct,** this narrow tube with a ciliated lining conducts a mature **ovum** from the area of the ovary toward the uterus. It is the usual site of fertilization. Long processes called **fimbriae** surround the tube's margin at its ovarian end.

☐ **3**   The **uterus,** or womb, is a single muscular sac at the midline in the pelvic cavity. Identify these parts of the uterus:
  ○ **body of the uterus**—the larger, rounded portion in which is located the main part of the **uterine cavity**
  ○ **cervix**—the narrow, inferior portion of the uterus (the **cervical canal** passes through this area to join the vagina)
  ○ **serous layer of the uterus**—the portion of the peritoneum that forms the outer uterine wall
  ○ **myometrium**—the muscular middle layer of the uterine wall
  ○ **endometrium**—the inner layer of the uterine wall

☐ **4**   The **vagina** is a muscular canal that receives the male penis during *sexual intercourse.* It functions as the **birth canal** during *delivery* and as an exit for sloughed-off endometrium during **menses.** The opening of the vagina to the outside may be covered partially or entirely by a thin, membranous **hymen.**

☐ **5**   The external genitals of the female are collectively known as the **vulva,** or **pudendum** (fig. 38.1). Identify these features associated with the vulva:
  ○ **vestibule**—the central space of the vulva, into which the vagina and urethra open (openings of the **vestibular gland** ducts conduct lubricating fluid into the vestibule)
  ○ **clitoris**—the female glans, a small, erectile structure in the anterior corner of the vestibule (the clitoris is covered with a fold of skin called the **foreskin,** or **prepuce**)
  ○ **labia minora** (sing., **labium minus**)—a pair of thin, longitudinal folds, one on each side of the vestibule
  ○ **labia majora** (sing., **labium majus**)—a pair of thick, longitudinal folds lateral to the labia minora (the space between the labia majora is called the **pudendal cleft**)
  ○ **mons pubis**—a fatty mound, covered with hair, over the pubic bone

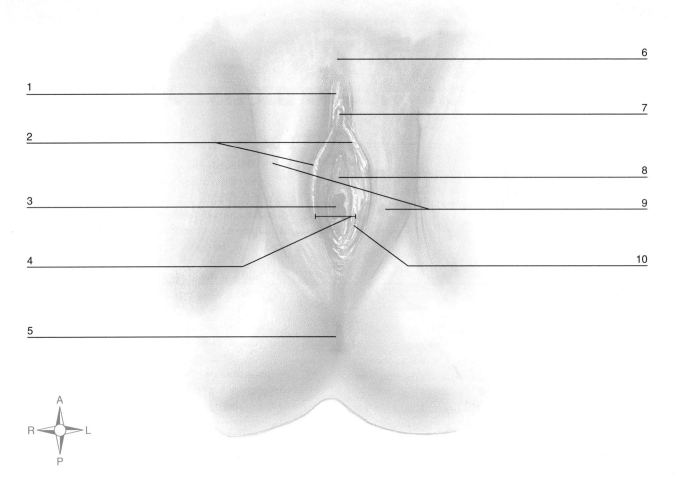

Figure 38.1   Label the indicated features of the vulva.

B. Microscopic Structure of the Ovary

First in a model or chart, then in a prepared microscopic specimen, locate important structures associated with the ovary (fig. 38.2).

SAFETY FIRST!

Be careful of electrical and other hazards when examining your specimen with a microscope.

☐ **1**   Each ovary is covered with a portion of the peritoneum. The outer region of each ovary is composed of dense connective tissue with numerous **ovarian follicles.** Each follicle contains an **oocyte,** the germ cell of the female.

☐ **2**   **Primary oocytes** are oocytes that have developed from **primordial follicles** to a point early in meiosis and then temporarily stopped developing. Each primary oocyte is surrounded by **granulosa cells.** The oocyte and granulosa cells together constitute a **primary follicle.**

☐ **3**   When stimulated by hormones, a primary follicle develops into a **secondary follicle.** Each secondary follicle has a central, fluid-filled **antrum** and a mass

of cells off to the side called the **cumulus mass.** The oocyte is embedded in the cumulus mass.

☐ **4**   The follicle eventually develops into a **mature follicle** (*Graafian follicle*) when the antrum fills with more fluid and the follicle becomes a bump on the ovary's surface. The tissue surrounding the follicle forms a layer called the **theca.** A clear layer, or **zona pellucida,** surrounds the enlarged oocyte. The developing follicle secretes the hormone **estrogen.**

☐ **5**   As more fluid fills the follicle, it bursts open and spills its contents into the peritoneal cavity near the fimbriae of the uterine tube. This process is called **ovulation.** The ruptured follicle then develops into the glandular **corpus luteum,** which secretes **progesterone** and estrogen. Later, the corpus luteum degeneratees to become the *corpus albicans.*

Hint... Figure 38.3 illustrates the structures discussed in this activity. Several ovarian follicles, each in different stages of development, may be found in a single specimen. However, their distribution is rather random. Refer to Histology Minireference Plate 3*b* for a micrograph of a primary follicle. Histology Minireference Plate 13*b* is a micrograph of a mature follicle.

The Female Reproductive System

OVARY 1
UTERINE TUBE 2

UTERUS
  BODY 3
  CERVIX 4
VAGINA 5
CLITORIS 6
LABIUM MINUS 7
LABIUM MAJUS 8
URINARY BLADDER 9
URETHRA 10
PUBIS 11
RECTUM 12

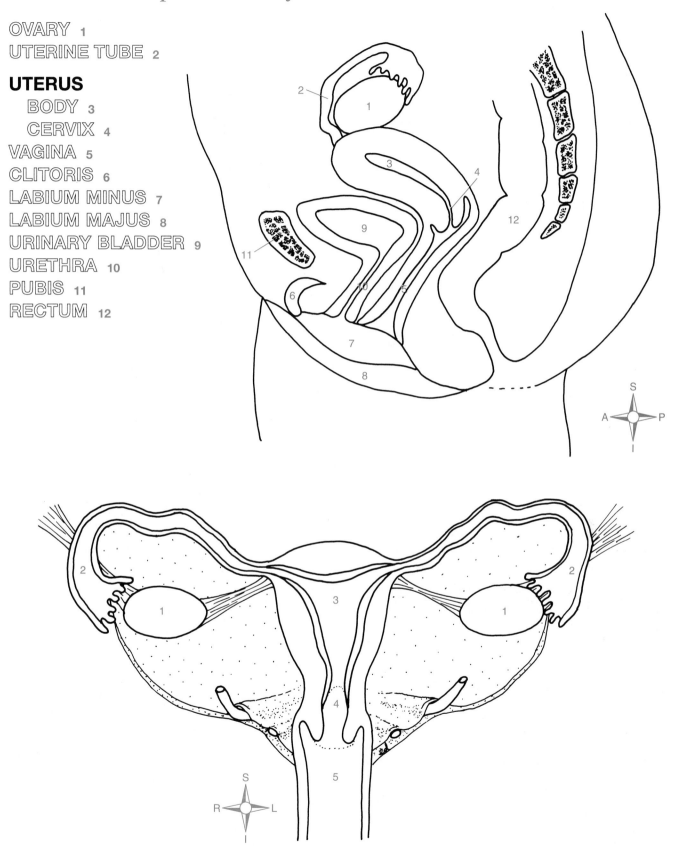

Figure 38.2 Color these features of the female reproductive system with colored pens or pencils.

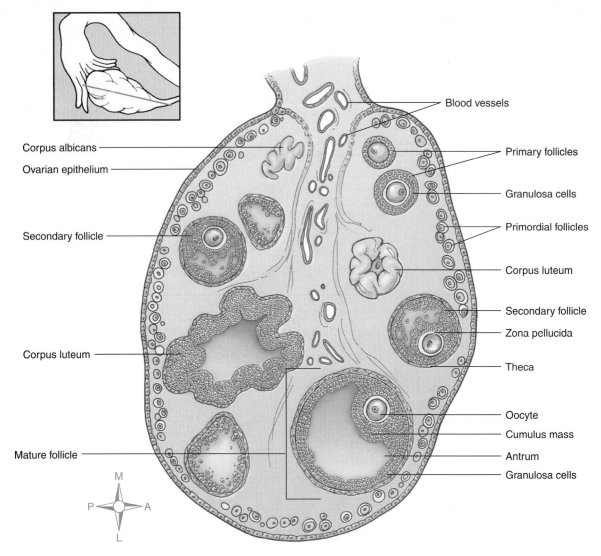

Blood vessels

Corpus albicans

Ovarian epithelium

Primary follicles

Granulosa cells

Primordial follicles

Secondary follicle

Corpus luteum

Secondary follicle

Zona pellucida

Corpus luteum

Theca

Oocyte

Cumulus mass

Mature follicle

Antrum

Granulosa cells

M

P — A

L

Figure 38.3 Sagittal section of the ovary, showing follicles in different stages of development. These stages do not all appear in the ovary at the same time. The follicles are drawn larger than actual scale within the ovary so they can be seen better.

Oogenesis

Oogenesis is the process by which ova, or egg cells, develop within tissue of the ovary. Figure 38.4 shows a flow chart of the main events of meiotic division that produce mature ova. Notice that the number of chromosomes of the original parent cells (oogonia, or *germ cells*) is reduced from 46 to 23 in the daughter ova cells. Notice also that the process involves two temporary pauses in the process of meiosis. The first pause begins just after the first meiotic division starts before birth, then restarts after puberty. The second pause begins just after the second meiotic division starts and does not resume until and unless a sperm cell unites with the ovum (*fertilization*). Notice also that, for every one oocyte that begins the process of oogenesis, only one viable daughter ovum results instead of the expected four daughter ova (compare with spermatogenesis, on p. 276). This occurs because at each division most of the cytoplasm remains with one of two daughter cells—thus ensuring that one will survive to the next stage and that the other will become a small *polar body* incapable of reproductive function. This means that the ovum will have a maximum amount of cytoplasm (containing nutrients) for successful development, rather than just an ordinary amount.

1. Oogonia are the cells from which oocytes arise. The oogonia divide by mitosis to produce other oogonia and primary oocytes.

2. Five million oocytes may be produced by the 4th month of prenatal life. They begin the first meiotic division but stop at prophase I. All of the primary oocytes remain in this state until puberty.

3. The first meiotic division is completed in a single mature follicle just before ovulation during each menstrual cycle. A secondary oocyte and the first polar body result from the unequal division of the cytoplasm.

4. The secondary oocyte begins the second meiotic division but stops at metaphase II.

5. The second meiotic division is completed after ovulation and after a sperm cell unites with the secondary oocyte. A secondary oocyte and a second polar body are formed.

6. Fertilization is completed after the nuclei of the secondary oocyte and the sperm cell unite. The resulting cell is called a zygote.

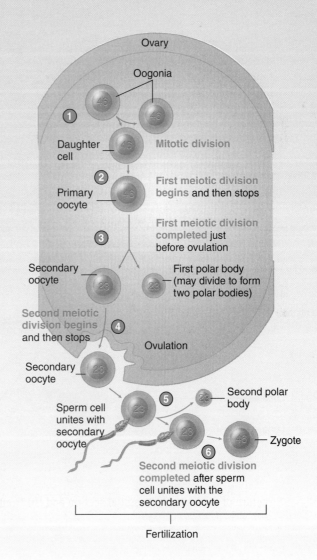

Figure 38.4 Oogenesis (showing meiotic division).

C. The Mammary Glands

Locate the female breast in a model or chart and figure 38.5. The **mammary glands** are modified sweat glands within the skin of the chest. Although structurally they are components of the integument, they are considered here because of their function of providing nutrition for offspring produced by the reproductive system. Find these features of the breasts:

☐ **1** Both the male and female breasts have a pigmented area of skin called the **areola.** The areola surrounds a raised **nipple.**

☐ **2** Normally, only the mammary glands of the female develop. At puberty, adipose tissue develops under the skin of the breast. This tissue supports 15 to 20 glandular **lobes** of the mammary gland. Each lobe has a single **lactiferous duct** that opens directly onto the surface of the nipple, giving the nipple a total of 15 to 20 openings.

☐ **3** Each lobe of the mammary gland is composed of several **lobules.** During *lactation* (milk production), the ends of the lobule ducts form secretory sacs called **alveoli.**

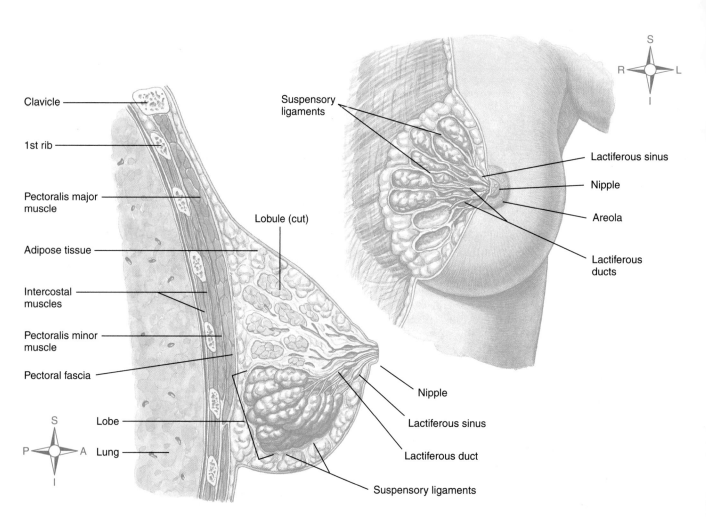

Figure 38.5 Structure of the breast: (*a*) sagittal section and (*b*) frontal view, breast dissected.

The Female Reproductive System

Name _____

Date _____ Section _____

Structure	Model or Chart	Function(s)
Ovary	☐	
Mesovarium	☐	
Broad ligament	☐	
Suspensory ligament	☐	
Ovarian ligament	☐	
Uterine tube	☐	
Fimbriae	☐	
Body of uterus, uterine cavity	☐	
Cervix, cervical canal	☐	
Serous layer of uterus	☐	
Myometrium	☐	
Endometrium	☐	
Vagina	☐	
Vestibule (of vulva)	☐	
Clitoris	☐	
Prepuce	☐	
Labium minus	☐	
Labium majus	☐	
Mons pubis	☐	
Ovarian follicles	☐	
Areola (of breast)	☐	
Nipple	☐	
Lobes (of mammary gland)	☐	
Lobules (of mammary gland)	☐	
Alveoli (of mammary gland)	☐	

Specimen: _____ *follicle*

Total magnification: _____

Figure 38.1 *(from p. 280)*

_____ 1

_____ 2

_____ 3

_____ 4

_____ 5

_____ 6

_____ 7

_____ 8

_____ 9

_____ 10

Fill-in

_____ 1

_____ 2

_____ 3

_____ 4

_____ 5

_____ 6

_____ 7

_____ 8

_____ 9

_____ 10

_____ 11

_____ 12

Fill-in

1. The fluid-filled space within a follicle is called the __?__.
2. The tissue surrounding a mature follicle is called the __?__.
3. A follicle secretes the hormone __?__.
4. The external genitals of the female are known by the term __?__.
5. The __?__ secrete lubricating fluid into the vestibule.
6. The release of an oocyte from a follicle is termed __?__.
7. The normal site of fertilization of an egg is in the __?__.
8. The narrow portion, or neck, of the uterus is called the __?__.
9. The oocyte in a secondary follicle is embedded in the __?__ mass.
10. The thin membrane covering all or part of the vaginal opening is called the __?__.
11. The __?__ is a structure of the ovary that secretes progesterone.
12. The birth canal is also called the __?__.

Development

Human development begins from the moment of fertilization in the female reproductive tract. A sperm cell from the male parent and an oocyte from the female parent unite to form the first cell of the offspring. This cell divides (using mitosis) again and again, forming the mass of cells that will eventually become a mature human. The first activity of this exercise presents a brief exploration of the stages of human prenatal development. The second activity offers a look at the circulatory plan of the fetus.

Before You Begin

☐ Read the appropriate chapter in your textbook.

☐ Set your learning goals. When you finish this exercise, you should be able to:
  o describe the basic plan of prenatal development
  o identify major features of successive developmental stages in models or charts
  o describe the fetal circulatory plan and the changes in circulation that occur around the time of birth

☐ Prepare your materials:
  o models or charts of early human development
  o *Life's Greatest Miracle* (Nova/PBS)
  o model or chart of human fetal circulation

☐ **Carefully** read the directions and safety tips for this exercise before starting any procedure.

A. Early Human Development

Obtain a set of models or charts that depict the major stages of early human development. Use the guidance given here and in your textbook to identify the major events of early development and the landmark structures associated with each stage.

☐ **1** The original cell of an offspring is termed the **zygote.** About 18 to 36 hours after being formed by the fusion of an egg and a sperm, this cell divides. The two daughter cells divide, then their daughter cells divide, and so on until a ball of cells is formed. Locate a zygote in figure 39.1 and in your model.

☐ **2** When the ball has about 32 cells and a fluid-filled cavity called the **blastocele** forms, the offspring is called a **blastocyst.** Most of the blastocyst is a single layer of cells called the **trophoblast,** but one portion also has an **inner cell mass** several cells thick. The inner cell mass develops into the embryo. The trophoblast develops into the **chorion,** which later becomes the placenta and the membranes that surround the embryo. Locate this stage in figure 39.1 and in your model.

☐ **3** Around 7 days after fertilization, the blastocyst begins **implantation** by digesting its way into the endometrium of the mother's uterus. During this stage, the chorion (formerly the trophoblast) develops projections (**chorionic villi**) into the maternal blood supply. This close association of the embryonic and maternal tissue is called the **placenta.** Later in development, the placenta allows diffusion of substances between the embryo's blood and the mother's blood. Identify this stage in figure 39.1, then in your model.

☐ **4** By day 11, several features have become visible. Identify these in figure 39.1, then in a model or chart:
  o **connecting stalk**—a narrow piece of tissue that connects the inner cell mass to the developing placenta (it will eventually develop into the **umbilical cord**)
  o **amniotic cavity**—a fluid-filled space within the cell mass, within a layer of cells called the **ectoderm** (this cavity will eventually develop into the **amniotic sac** surrounding the embryo)
  o **yolk sac**—another fluid-filled space within the cell mass, this one within a layer of cells called the **endoderm**
  o **embryonic disk**—a flat sheet of tissue formed by adjacent layers of endoderm and ectoderm

☐ **5** By day 14, more features of the developing offspring become apparent, as seen in figure 39.1:
  o **primitive streak**—a thickened line in the endoderm, at the center of the elongated embryonic disk (the embryo will eventually form around the streak; the **notochord** extends from the cephalic end of the streak)
  o **mesoderm**—a new *germ layer* that arises between the ectoderm and mesoderm

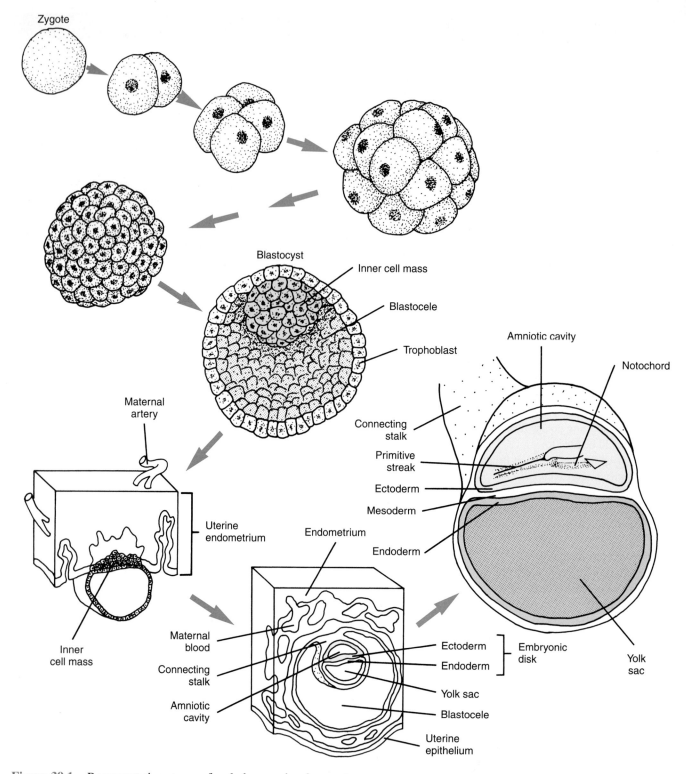

Figure 39.1 Representative stages of early human development.

☐ **6** Follow the developmental stages represented in figure 39.2 and in your model or chart, noting that the three germ layers (endoderm, mesoderm, and ectoderm) develop into different tissues and organs. Using your textbook or a reference book, determine which major tissues or organs are derived from each of the three germ layers. Report your findings in Lab Report 39.

☐ **7** After about 60 days, the cartilage and membranous tissue of the embryonic skeleton begin to ossify.

Figure 39.2 Embryonic and fetal development (sketches are drawn at different scales).

From this point onward, the offspring is termed a **fetus.** Examine a model or chart of a full-term fetus. Because its organs and systems have developed sufficiently, the full-term fetus is able to survive outside the uterus. Locate these structures associated with the fetus:

- **Amniotic sac**—This is a fluid-filled space derived from the amniotic cavity. It cushions the developing offspring. At **parturition,** the sac breaks and spills its contents through the birth canal (vagina).
- **Placenta**—This organ is originally formed by the chorionic villi's implantation into the endometrial lining of the uterus. The mature placenta allows the exchange of materials between the fetal circulatory system and the maternal circulatory system. Examine the anatomical relationship of these two sets of vessels.
- **Umbilical cord**—This is derived from the connecting stalk observed in earlier stages. The umbilical cord includes **umbilical arteries** and an **umbilical vein** (fig. 39.3). Activity B describes how these vessels fit into the fetal circulatory plan. The umbilical cord serves as a connection between the developing offspring and the mother's body.

☐ **8** When your exploration of the models and charts is complete, view the film *Life's Greatest Miracle,* if available. This film follows the course of human development from fertilization onward and is an outstanding summary activity for your study of human reproduction and development.

B. Fetal Circulation

Some of the many dramatic anatomical and physiological changes that occur at parturition are the changes in the cardiovascular circulation. The fetal circulatory plan functions well in the circumstances found *in utero* but must change quickly if a newborn is to survive outside the uterus. Examine figure 39.3, or a model or chart of fetal circulation, and note these special features:

☐ **1** The **placenta** has already been described. Located on the inside wall of the uterus, one function of the placenta is the exchange of nutrients and wastes between fetal and maternal blood. Most materials move by means of diffusion, crossing the membranes that separate the two blood systems.

☐ **2** The **umbilical vein** conducts oxygenated fetal blood from the placenta, through the umbilical cord, to and through the liver, and into the **ductus venosus.** The ductus venosus drains into the inferior vena cava, which returns the blood to the fetal heart (right atrium).

☐ **3** From the right atrium, oxygenated blood may flow through the **foramen ovale** into the left atrium. This effectively bypasses the entire pulmonary loop. Some right atrial blood moves into the right ventricle. Right ventricular blood exits the heart by way of the pulmonary artery. Near the branching of the pulmonary artery into left and right branches, the **ductus arteriosus** provides another bypass of the pulmonary loop. Blood tends to bypass circulation to the lungs here and at the right atrium because of the high resistance to blood flow in the developing lungs. Only a small proportion of blood reaches the lungs, just enough for its continued development.

☐ **4** Blood in the systemic circulation may enter the two **umbilical arteries** that branch from the internal iliac arteries. The umbilical arteries conduct partially deoxygenated blood through the umbilical cord to the placenta. Follow the possible circulatory paths through the fetal body several times, noting the differences between fetal and adult patterns of blood circulation.

☐ **5** Your model or chart may indicate the changes that occur at parturition to shift the fetal circulatory pattern to the adult pattern. The major changes are noted here:

- Flow through the umbilical arteries and vein stops when the umbilical cord is cut and tied, and the vessels degenerate. The umbilical vein and ductus venosus are replaced by ligaments.
- When the lungs expand during the respirations that begin after birth, pulmonary resistance to blood flow is reduced. Therefore, blood tends to flow through the pulmonary loop and tends not to flow through the foramen ovale or into the ductus arteriosus. Increased left atrial pressure forces a flap of the septum over the foramen ovale, which eventually seals the opening completely. The ductus arteriosus constricts and is later replaced by a fibrous ligament.

Ascending aorta

Superior vena cava

Right atrium

Foramen ovale

Inferior vena cava

Right ventricle

Ductus venosus

Liver

Portal vein

Umbilical vein

Umbilical cord

Fetal umbilicus

Placenta

Umbilical arteries

Ductus arteriosus

Pulmonary artery

Pulmonary veins

Left atrium

Thoracic aorta

Left ventricle

Abdominal aorta

Hepatic artery

Kidney

Renal vein and artery

Intestine

Left common iliac artery

External iliac artery

Internal iliac (hypogastric) arteries

Figure 39.3 The fetal circulatory plan.

Ultrasonography

Ultrasonography is a method of imaging body parts by using high-frequency sound (ultrasound) waves. Usually, a handheld wand that emits ultrasonic waves is placed against the patient's skin. The waves pass through different tissues at different speeds, and some waves are reflected back to the wand. Sensors in the wand detect the reflected ultrasound and relay the information to a computer capable of constructing an image based on the reflections. The image produced by this technique is called a **sonogram** (fig. 39.4).

An advantage of the technique is that ultrasound energy is not as dangerous as the radiation used in CT scans and regular radiographs. This is one reason that it has become popular among obstetric physicians. Developing fetal tissue that could be harmed by x-rays is not harmed by ultrasound waves. Therefore, the development of an embryo or a fetus can be monitored by ultrasonography.

Sonograms can be used to determine whether there is a single fetus or multiple fetuses. Abnormally developed structures are sometimes visible. Late in development, male or female genitals can sometimes be seen.

Occasionally, measurements of a fetus at various stages are made. The normal rate of development of organs can then be verified. Usually, the age of a fetus is given as the **clinical age** rather than the **developmental age** used earlier in this exercise. The clinical age is the time since the mother's last menstrual period, whereas the developmental age is the time since fertilization.

(a)

(b)

Figure 39.4 (*a*) This sonogram shows the outline of a fetal skull at 153 days (clinical age). Compare it with (*b*) the skull of the same fetus at 206 days.

1. The size of the skull has increased by what percentage during the 53 days between two sonograms shown in fig. 39.4?

2. Which structures in particular do you think are likely to show up well in a sonogram?

3. What other applications for ultrasonography can you think of?

4. What advantages do CT scans, MR images, and regular radiographs have over sonograms?

LAB REPORT 39

Development

Name _____

Date _____ Section _____

Identify

_____ 1

_____ 2

_____ 3

_____ 4

_____ 5

_____ 6

_____ 7

_____ 8

_____ 9

_____ 10

_____ 11

_____ 12

_____ 13

_____ 14

_____ 15

_____ 16

_____ 17

_____ 18

_____ 19

_____ 20

_____ 21

_____ 22

_____ 23

_____ 24

_____ 25

_____ 26

_____ 27

Identify

Identify the germ layer from which each of these organs or tissues is derived. Write *endo* for endoderm, *ecto* for ectoderm, and *meso* for mesoderm.

1. adrenal medulla
2. anterior pituitary gland
3. bones of the face
4. bones (except those of the face)
5. brain
6. cardiovascular organs
7. dermis
8. epidermis
9. gonads
10. kidney ducts and bladder
11. kidneys
12. lens and cornea of eye
13. lining of GI tract
14. linings of hepatic and pancreatic ducts
15. lining of lungs
16. melanocytes
17. muscle
18. nasal cavity
19. outer ear
20. parathyroid gland
21. skeletal muscles of the head
22. spinal cord
23. thymus gland
24. thyroid gland
25. tonsils
26. tooth dentin and pulp
27. tooth enamel

Multiple choice

_____ 1

_____ 2

_____ 3

_____ 4

_____ 5

_____ 6

_____ 7

_____ 8

Multiple choice

1. The cell formed by the union of a sperm cell and an oocyte is called
    - a. a blastula.
    - b. a morula.
    - c. a zygote.
    - d. a blastocele.
    - e. *b* and *c*.

2. The connecting stalk seen early in human development later becomes the
    - a. placenta.
    - b. amniotic sac.
    - c. umbilical cord.
    - d. yolk sac.

3. The earliest developmental stage at which one can see a fluid-filled extracellular cavity is the
    - a. morula.
    - b. blastula.
    - c. zygote.
    - d. fetus.
    - e. four-cell stage.

4. The organ in which maternal and fetal blood exchange material is the
    - a. umbilical cord.
    - b. umbilical vein.
    - c. amniotic sac.
    - d. amniotic cavity.
    - e. placenta.

5. The process by which a baby is born is called
    - a. gestation.
    - b. pregnancy.
    - c. differentiation.
    - d. parturition.
    - e. lactation.

6. The chorion is derived from the
    - a. trophoblast.
    - b. blastocele.
    - c. inner cell mass.

7. The embryo develops around a thickened line in the endoderm called the
    - a. trophoblast.
    - b. inner cell mass.
    - c. notochord.
    - d. primitive streak.

8. In the fetus, right atrial blood may flow into
    - a. the right ventricle.
    - b. the left ventricle.
    - c. the left atrium.
    - d. the ductus venosus.
    - e. *a* and *c*.

Genetics and Heredity

Heredity, or the inheritance of traits, falls into the realm of **genetics.** Genetic information, in the form of **genes** found in the 23 DNA molecules that you inherit from your mother and the 23 that you inherit from your father, is passed from generation to generation. Twenty-two pairs of your DNA molecules, or *chromosomes,* are called **autosomes.** The remaining pair are called the **sex chromosomes.** One member of each pair is inherited from one parent, the other member of the pair from the other parent.

Each autosome in a pair is homologous to the other member of the pair. With this arrangement, a person has two genes for every inherited characteristic. If one gene is always expressed, whether or not its mate is the same gene, geneticists call that gene **dominant.** A gene that is not expressed when its mate is different is termed a **recessive** gene.

All of the genes in all of the chromosomes together are called the **genome.** The study of all the genes in the human species, or the *human genome,* is called **genomics.**

Before You Begin

☐ Read the appropriate chapter in your textbook.

☐ Set your learning goals. When you finish this exercise, you should be able to:
  ○ identify the phenotype and possible genotypes of selected human characteristics
  ○ distinguish between the concepts of dominant and recessive as applied to inherited traits
  ○ predict probabilities of phenotypes and genotypes in offspring, given the parents' genotypes or phenotypes

☐ Prepare your materials:
  ○ taste papers: *control, sodium benzoate, PTC, thiourea*
  ○ BIOHAZARD container
  ○ blood typing materials (if not done previously in Lab Exercise 25)

☐ **Carefully** read the directions and safety tips for this exercise before starting any procedure.

A. Human Phenotypes and Genotypes

For any inherited characteristic, a **phenotype** can be identified. The phenotype is the characteristic actually expressed in an individual. A person's **genotype** is a statement of both genes that influence a particular trait. For example, *albinism* is a recessive genetic condition in which a person lacks skin pigmentation. By convention, a gene for normal skin pigmentation, which is dominant, is represented as *A*. The recessive gene is represented as *a*. A person with normal skin pigmentation may have the genotype *AA,* meaning that both genes are normal. Geneticists may state that the person's genotype is **homozygous,** meaning that both genes are the same. A person with normal skin color could also have the *Aa* genotype. Because *A* is dominant, the abnormal *a* is not expressed. This person has a **heterozygous** genotype, meaning that the two genes are not the same. A person with albinism must have the genotype *aa,* because that is the only genotype that allows the recessive abnormal trait to be expressed. Thus, the normal phenotype (normal skin color) may be associated with the genotype *AA* or the genotype *Aa.* The abnormal phenotype (albinism) is always associated with the genotype *aa.*

> *Hint . . .* Skin color is determined by a number of factors, both environmental and genetic. A number of different *A* genes exist in the human gene pool, and genes at other locations in the DNA may have an impact on skin color. The example of albinism is simple when first observed, but the influence of multiple factors complicates things. This is typical of inherited characteristics. Often, many genes influence one particular trait. Also, many environmental factors can affect inherited characteristics.

This activity instructs you to determine your phenotype for a variety of easily observed characteristics. Once you know your phenotype, you can determine your possible genotypes. For example, if you are an albino, you know that your genotype is *aa.* If you do not have albinism, your genotype is either *AA* or *Aa.* For each characteristic described, record your phenotype and possible genotypes in Lab Report 40.

☐ **1** The first set of characteristics to be determined involves the sense of taste. It is known that the ability to taste certain compounds depends on the presence of certain genes. You will taste papers that have each been impregnated with a different compound. Before starting, put a piece of CONTROL test paper on your tongue and chew it. If you taste something, you must be sensitive to a compound in the paper itself. Ignore that taste in the tests that remain, or try to sense *differences* in taste.

   ○ **Sodium benzoate test**—The ability to taste something sweet, salty, or bitter in the paper is dominant.
   ○ **PTC (phenylthiocarbamide) test**—The ability to sense a bitter taste is dominant.
   ○ **Thiourea test**—The ability to taste something bitter is dominant.

☐ **2** The next set of determinations involves anatomical characteristics of your hand (fig. 40.1):

   ○ **Bent little finger**—Place your relaxed hand flat on the lab table. If the distal phalanx of the little (fifth) finger bends toward the fourth finger, you have the dominant trait.
   ○ **Middigital hair**—Dorsal hair on the skin over the middle phalanges of the hand is dominant.
   ○ **Hitchhiker's thumb**—If you can hyperextend the distal joint of the thumb noticeably, you have the recessive trait.

☐ **3** Determine your phenotype for these facial features:

   ○ **Pigmented anterior of the iris**—If you have pigment on the anterior *and* posterior of the iris, your eyes are green, brown, black, or hazel. If you lack pigment on the anterior aspect of the iris, your eyes are blue or gray. Anterior pigmentation is dominant.
   ○ **Attached earlobes**—If the inferior, fatty lobe of the ear is attached rather than free, you have the recessive trait (fig. 40.2).
   ○ **Widow's peak**—Assuming you have a hairline, if it is straight across the forehead, you have the recessive trait. If it forms a downward point near the midline, you have the dominant widow's peak (fig. 40.2b).
   ○ **Tongue roll**—Try to curl your tongue as you extend it from your mouth. If you can't curl it, you have the recessive trait (fig. 40.2c).

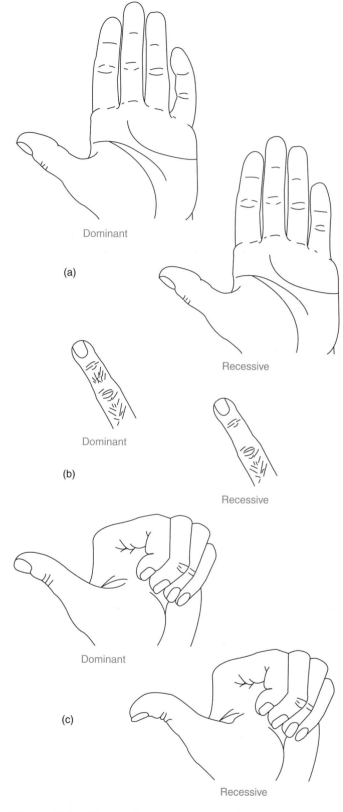

Figure 40.1 The dominant and recessive forms of (*a*) bent little finger, (*b*) middigital hair, and (*c*) hitchhiker's thumb.

(a)   Dominant   Recessive

(b)   Dominant   Recessive

(c)   Dominant   Recessive

Figure 40.2 Dominant and recessive forms of (*a*) attached earlobes, (*b*) widow's peak, and (*c*) tongue roll.

o **Freckles**—If your face has a scattering of freckles, you have the dominant form of this characteristic. If your face is free of freckles, you have the recessive condition.

☐ **4** There are two dominant genes for ABO blood types. One is I^A, which signifies the presence of the A antigen. The other is I^B, signifying the B antigen. The recessive gene is i, signifying neither ABO antigen. A person with the phenotype type A has the genotype $I^A I^A$ or $I^A i$. A person with the phenotype type B has the genotype $I^B I^B$ or $I^B i$. A person with phenotype type AB has the genotype $I^A I^B$. Phenotype type O requires the genotype ii. Record your phenotype (your ABO blood type). What is your genotype?

Hint . . . If you have not already typed your blood (as instructed in Lab Exercise 25), consult your health records or perform the typing now. Refer to Lab Exercise 25 and heed the safety advice given there.

☐ **5** The Rh blood type is also determined by genetics. Presence of the Rh antigen is dominant. What is your Rh blood type? What is your genotype?

B. Probabilities of Inheritance

Now that you have a grasp of the concept of phenotype and genotype, you can move on to the concept of **probability.** Probability is the likelihood of a certain outcome in a particular event. As applied to human genetics, *probability* refers to the likelihood that the offspring of a particular set of parents will have a certain inherited condition. **Genetic counselors** work with prospective parents to determine their possible genotypes for a variety of traits. Then they predict the probability of their children having those traits. In this way, parents can anticipate possible abnormal genetic conditions.

In this exercise, you will use a simple method developed by the English geneticist Punnett. He devised a simple grid, or **Punnett square,** with which one can easily predict simple ratios of genetic probability.

First, the following is an example:

*The father has freckles; the mother does not. The gene for freckles is **F**, and the gene for no freckles is **f**. The father's genotype is either **FF** or **Ff**. The mother's genotype must be **ff**. We must consider two different outcomes in offspring from this couple because the father has two possible genotypes for this trait. First, let's assume that the father's genotype is **FF**. Any sperm cell contributed by the father will have the **F** gene. Any oocyte contributed by the mother will have the **f** gene. We place the possible gene for the father in the left margin of the Punnett square and the mother's in the top margin:*

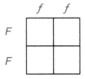

To use the Punnett square, start in the top left square of the grid. Combine the gene from the left margin with the gene from the top margin:

We now have one possible offspring genotype: **Ff.** We then do the same for the other three squares of the grid:

Our Punnett square predicts a 100% probability that any one offspring will have genotype **Ff** and therefore have the freckled phenotype. The second possibility is that the father's genotype is **Ff.** We can set up a Punnett square for this possibility:

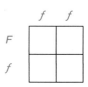

We can now fill in the grid with possible offspring genotypes, combining the gene from the left margin with the gene from the top margin for each square:

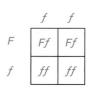

Because half the squares have genotype **Ff,** there is a 50% probability that any offspring of this couple will have the **Ff** genotype (and therefore the freckled phenotype). There is also a 50% probability that the genotype will be **ff,** expressed as the nonfreckled phenotype. If the phenotypes or genotypes of the clients' parents and grandparents are known, a genetic counselor might construct a **pedigree,** or family tree, to determine the father's likely genotype. If that information is unavailable, the parents will have to deal with the two different probabilities.

Now that you have seen an example, try your hand at the following problems. Draw your Punnett squares and record your results in Lab Report 40.

☐ **1** Huntington's disease (HD), or Huntington's chorea, is a degenerative nerve disorder with a genetic basis that becomes apparent after about the age 40. The abnormal gene that produces this disease, *H,* is dominant. The normal, recessive gene is *h.* One of Heather's parents has HD but the other doesn't. Can you predict the highest probability that Heather will develop HD later in her life?

☐ **2** Kevin has Rh-positive blood. His wife, Christine, has Rh-negative blood. Their first child, Andrew, has Rh-positive blood. Both of Kevin's parents have Rh-positive blood. What is the probability that the child Kevin and Christine are expecting will be Rh-negative?

☐ **3** Leo's father has albinism, but Leo doesn't. Cleo's father has albinism, but she doesn't. If Leo and Cleo have a child, what is the probability that it will have albinism? What is the probability that their second child will have albinism? Their third child?

☐ **4** In the ABO blood typing system, Mario is type O. Ana is type AB. What ABO blood types might their children have?

LAB REPORT 40

Genetics and Heredity

Name _____

Date _____ Section _____

Trait	Dominant Gene(s)	Recessive Gene	Your Phenotype	Possible Genotypes
Sodium benzoate taste	S	s		
PTC taste	P	p		
Thiourea taste	T	t		
Bent little finger	L	l		
Middigital hair	M	m		
Hitchhiker's thumb	H	h		
Pigmented anterior of iris	I	i		
Attached earlobes	A	a		
Widow's peak	W	w		
Tongue roll	R	r		
Freckles	F	f		
ABO blood type	I^A, I^B	i		
Rh blood type	D	d		

☐ 1.   What is the highest probability that Heather will develop Huntington's disease? _____%

☐ 2.   What is the probability that Kevin and Christine's second child will have Rh-negative blood? _____%

☐ 3.   What is the probability that Leo and Cleo's first child will be albino? _____%

Their second child? _____%

Their third child? _____%

☐ 4.   What are the possible ABO blood types of Mario and Ana's children? _____%

The Laboratory Rat

This exercise may be used as either an introduction to vertebrate anatomy or a synthesis activity to conclude the laboratory course. The previous exercises in this lab manual have concentrated on organs or systems one by one. This exercise draws everything together in an exploration of the entire vertebrate body. Because the vertebrate body plan is similar among species, another laboratory animal specimen (e.g., a fetal pig) could be substituted.

Hint . . . In a short course, a well-preserved specimen can be used from time to time throughout your studies. If you will be looking at your dissected specimen from time to time for the next several months, make sure that it is kept in the appropriate container under conditions suggested by your instructor.

Before You Begin

☐ Set your learning goals. When you finish this exercise, you should be able to:
  o perform a whole-body dissection of a vertebrate animal
  o identify the major anatomical features of the vertebrate body in a dissected specimen

☐ Prepare your materials:
  o preserved (plain or double-injected) or freshly killed laboratory rat (or similar vertebrate)
  o dissection tools and trays
  o mounted rat skeleton (optional)
  o storage container (if preserved specimen is to be reused)

☐ **Carefully** read the directions and safety tips for this exercise before starting any procedure.

SAFETY FIRST!

Observe the usual precautions when working with a preserved or fresh specimen. Heed the safety advice accompanying preservatives used with your specimen. Use protective gloves and safety eyewear while handling your specimen. Avoid injury with dissection tools. Dispose of your specimen as instructed.

A. The External Aspect

Examine the external aspect of your specimen:

☐ 1 Determine the anatomical orientation of the specimen. Which direction is anterior? Posterior? Which direction is ventral? Dorsal? Identify sagittal, transverse, and frontal planes in your specimen.

☐ 2 Identify these externally visible features:
  o **pinna (auricle)**
  o **external nares (nostrils)**
  o **vibrissae (whiskers)**
  o **incisors**
  o **integument**
  o **forelimbs**
  o **hindlimbs**
  o **thoracic region**
  o **abdominal region**
  o **nipples**
  o **anus**
  o **tail**

☐ 3 Determine the sex of your specimen by examining the external genitals:
  o **Female**—Immediately anterior to the anus, on the ventral surface, is the **vulva** with an opening to the **vagina.** Anterior to the vulva is the **clitoris** with the **urethral** opening.
  o **Male**—The **scrotum** containing the **testis** is immediately anterior to the anus, perhaps even hiding the anus from view. Near the anterior edge of the scrotum is the **penis** with its **prepuce,** or skinfold covering. Locate the opening of the **urethra** in the penis.

Hint . . . A complete set of anatomical sketches of the rat is found in the "Anatomical Atlas of the Rat" in this exercise (figs. 41.1–41.7).

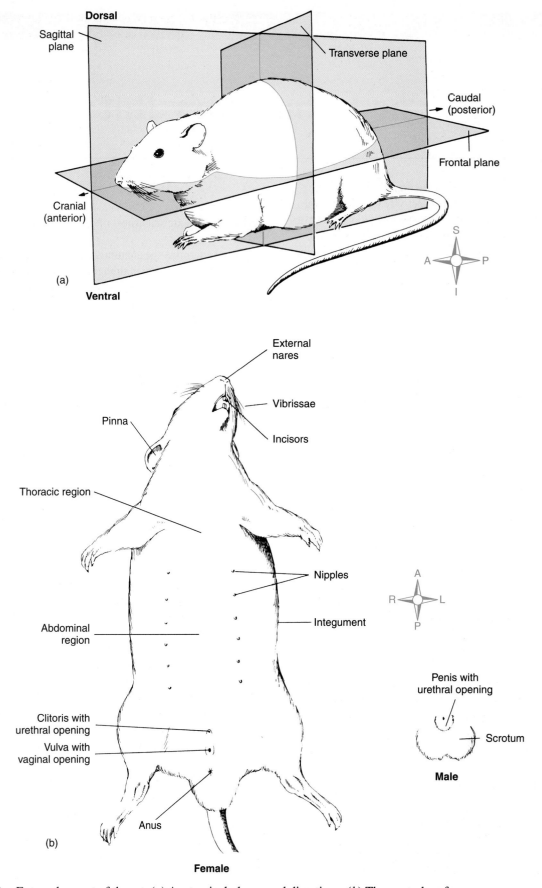

Dorsal

Sagittal plane

Transverse plane

Caudal (posterior)

Frontal plane

Cranial (anterior)

(a)

Ventral

External nares

Vibrissae

Incisors

Pinna

Thoracic region

Nipples

Integument

Abdominal region

Clitoris with urethral opening

Vulva with vaginal opening

Anus

(b)

Female

Penis with urethral opening

Scrotum

Male

Figure 41.1 External aspect of the rat. (*a*) Anatomical planes and directions. (*b*) The ventral surface.

B. Skin, Bones, and Muscles

☐ **1** Remove most of the skin from your specimen by
following this procedure:
   o Place the animal in the tray with its ventral
   surface facing you.
   o Pull up on the skin over the sternum and
   puncture it with the tip of a scissors. Slide the
   bottom tip of the scissors into the **subcutaneous**
   area under the skin.
   o Begin cutting along the lines indicated in
   figure 41.2. Be careful not to cut into the
   skeletal muscles under the skin.
   o With your forceps, pull the two flaps of skin over
   the neck away from the animal's body. Notice
   the **areolar tissue** under the skin that is pulled
   apart as you remove the skin. Pull the flaps of
   skin over the abdomen and over the **groin** area
   away in a similar fashion.
   o Turn the animal over so that the dorsum is
   facing you.
   o If you made all the cuts properly, you should be
   able to pull the skin off the back in one complete
   piece. Sometimes it helps if you scrape at the
   loose connective tissue under the skin with your
   scalpel as you peel the skin away. The skinning
   process is difficult unless you have patience and
   proceed slowly.

☐ **2** Examine the skin, identifying these features:
   o **dermis**—the thick, inner layer of the skin
   o **epidermis**—the thinner, outer layer of the skin
   with **hair** (fur)

☐ **3** Explore the shape of the skinned rat body. How many
bones of the rat's skeleton can you see or palpate
(feel)? If you have a mounted rat skeleton available,
identify as many of the bones of the skeleton as you
can. If you become stumped, refer to figure 41.3a.
Notice the similarity between the rat's skeletal plan
and that of the human.

☐ **4** Observe the rat's musculature. Some of the external
muscles of the torso can be separated from each other
for easier viewing. Slide a probe into the loose
connective tissue joining adjacent muscles and run
the probe along their margins. Using figures 41.3 and
41.4 as guides, try to identify the major superficial
muscles of the rat's body.

C. Cardiovascular Structures

☐ **1** Open the **ventral body cavity** by cutting into its
muscular wall in a manner similar to your earlier cut
into the skin. Cut flaps in the neck, abdomen, and
groin areas as you did with the skin, but do not cut all
the way around to the back. Be careful not to damage
any **visceral organs** with your scissors as you cut.
Fold back the flaps and anchor them with pins or
remove them.

☐ **2** Locate the **heart,** near the middle of the **thoracic
cavity,** and the **mediastinum.** Can you identify the
four chambers?

☐ **3** If you have a double-injected preserved specimen,
the **arteries** are filled with red latex and the **veins** are
filled with blue latex. If not, the arteries can usually
be distinguished from veins because they are stiffer
and lighter in color than veins. Locate the **aorta,** the
large artery leaving the heart and arching posteriorly.
Trace the branches of the aorta, naming them if you
can. Use figure 41.5a if you need help.

☐ **4** Locate the **anterior vena cava** where it drains into
the heart. This is analogous to the superior vena cava
in the human. Follow its tributary veins, and identify
them with the help of figure 41.5b. Locate the
posterior vena cava and trace its tributaries.

Hint . . . Once you have cut into the ventral body
cavity, you may be tempted to cut and remove organs. It is
important that you keep everything as intact as possible. You
may pull organs to one side or another to view deeper
structures, but avoid making cuts.

D. The Viscera

The viscera, or major internal organs, can be seen within the
ventral body cavity. Use figures 41.6 and 41.7 to guide you
in locating the following:

☐ **1** Locate some of these features of the lower
respiratory system:
   o **larynx**
   o **trachea**
   o **primary bronchi**
   o **lungs** (can you distinguish the parietal and
   visceral **pleurae?**)
   o **diaphragm**

☐ **2** Locate these structures of the **digestive system:**
   o **submandibular salivary glands**
   o **esophagus**
   o **stomach**
   o **liver** (the gallbladder is absent in the rat)
   o **pancreas**
   o **small intestine**
   o **mesentery**
   o **cecum** (this structure is very large in the rat and
   does not have an appendix)
   o **colon**

☐ **3** Locate these **lymphatic organs:**
- spleen
- **thymus** (the thymus may be small in older animals)

☐ **4** Locate these features of the **urinary system:**
- **kidney**
- **ureter**
- **urinary bladder**
- **urethra**

☐ **5** Try to locate these **endocrine glands** in your specimen:
- **thyroid gland**
- **thymus gland**
- **pancreas**
- **adrenal glands**
- **testes**
- **ovaries**

☐ **6** Identify these structures associated with the male **reproductive system:**
- **testes**
- **epididymis**
- **ductus (vas) deferens**
- **seminal vesicle**
- **prostate gland**
- **penis**

☐ **7** Find these **female reproductive system** structures:
- **ovaries**
- **uterine tubes**
- **uterus** (notice that the rat uterus has a Y shape, with a right and left *uterine horn*)
- **vagina**

Hint . . . Unless your lab group has both a male and a female specimen, you may want to temporarily trade specimens with a group that has a rat of a different gender than yours. By doing so, you will be able to find the features of both reproductive systems.

☐ **8** If your specimen is a pregnant female, open the uterus *very carefully* and identify these structures:
- **Amniotic sac**—This is a fluid-filled cushion for each embryo.
- **Embryo**—Examine the structure of the embryo. Do you think it is in a late stage or an early stage of development?
- **Umbilical cord**—Can you distinguish among the umbilical vessels?
- **Placenta**—It is attached to the uterine lining.

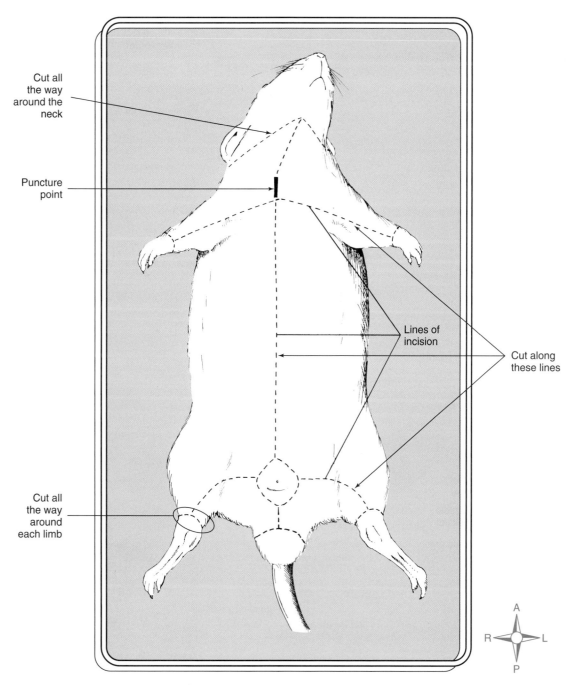

Cut all the way around the neck

Puncture point

Lines of incision

Cut along these lines

Cut all the way around each limb

A

R — L

P

Figure 41.2 Directions for cutting the skin of the rat.

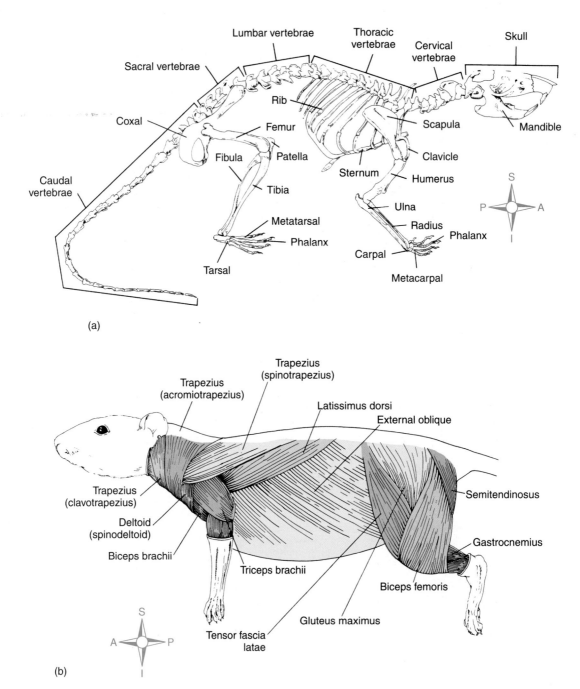

Figure 41.3 Lateral views of the rat. (*a*) The skeleton. (*b*) Superficial skeletal muscles.

Figure 41.4 Superficial skeletal muscles of the rat. (*a*) Dorsal aspect. (*b*) Ventral aspect.

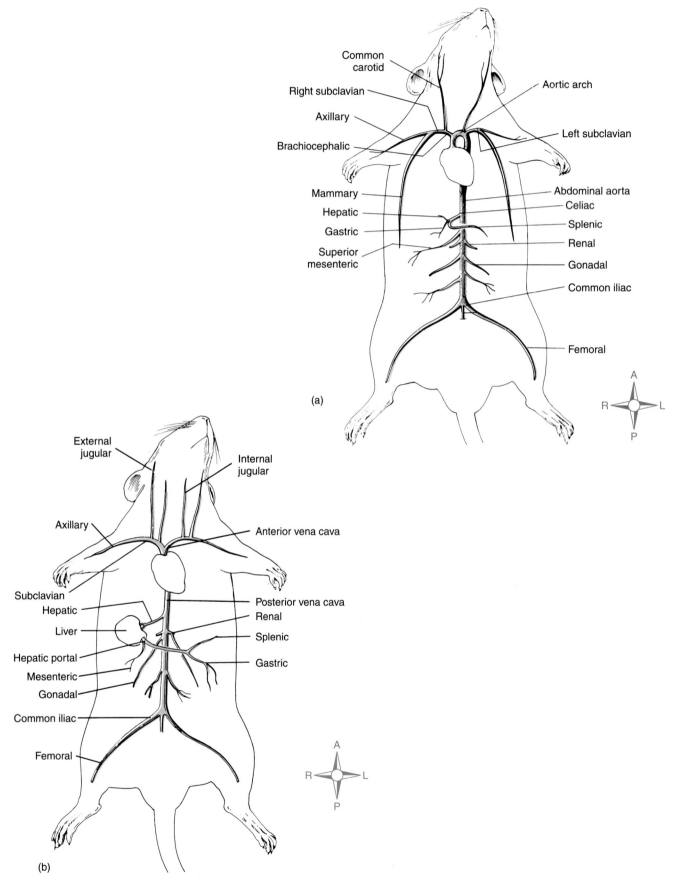

Figure 41.5 The rat's systemic circulation. (*a*) Major systemic arteries. (*b*) Major systemic veins.

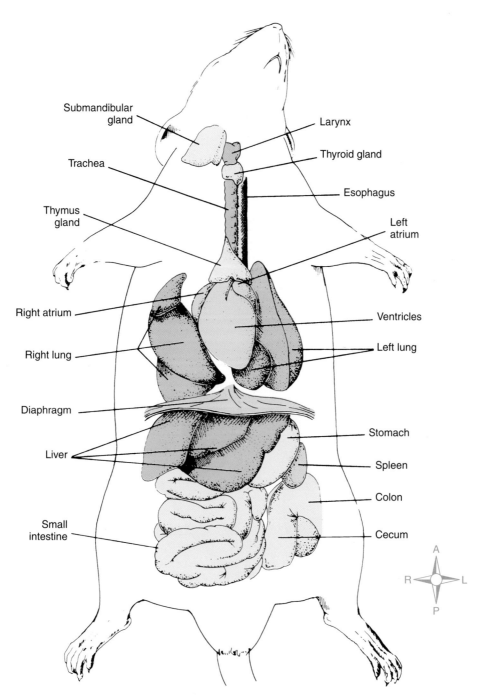

Figure 41.6   Organs of the ventral body cavity of the rat.

(a)

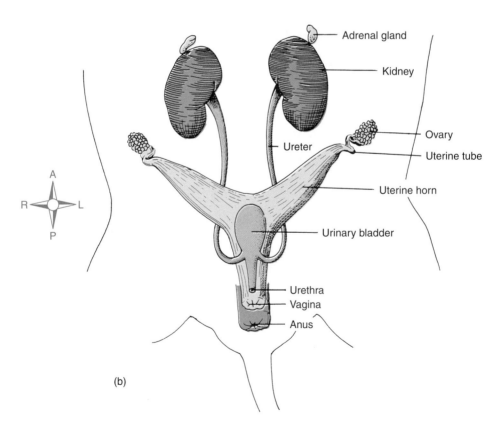

(b)

Figure 41.7 Organs of the pelvic cavity of the rat. (*a*) Male. (*b*) Female.

The Laboratory Rat

Name _____

Date _____   Section _____

Rat Dissection Checklist

□ **External Aspect**

□ Pinna (auricle)

□ External naris (nostril)

□ Vibrissae (whiskers)

□ Incisors

□ Integument

□ Forelimbs

□ Hindlimbs

□ Thoracic region

□ Abdominal region

□ Nipples

□ Anus

□ Tail

Gender: □ female

□ Vulva

□ Vagina

□ Clitoris

□ Urethral opening

Gender: □ male

□ Scrotum (□ testes within)

□ Penis

□ Prepuce (foreskin)

□ Urethral opening

□ **Skin, Bones, and Muscles**

□ Subcutaneous areolar tissue

□ Dermis

□ Epidermis, fur

□ Axial skeleton

□ Cranial bones

□ Facial bones

□ Cervical vertebrae

□ Thoracic vertebrae

□ Lumbar vertebrae

□ Sacral vertebrae

□ Caudal vertebrae

□ Ribs

□ Sternum

□ Appendicular skeleton

□ Scapula

□ Clavicle

□ Humerus

□ Radius

□ Ulna

□ Carpal bones

□ Metacarpal bones

□ Phalanges of the forelimb

□ Coxal bone

□ Femur

□ Patella

□ Tibia

□ Fibula

□ Tarsal bones

□ Metatarsal bones

□ Phalanges of the hindlimb

□ Head, neck, and shoulder muscles

□ Sternomastoid (human: sternocleidomastoid)

□ Clavotrapezius (human: trapezius)

□ Acromiotrapezius (human: trapezius)

□ Spinotrapezius (human: trapezius)

□ Spinodeltoid (human: deltoid)

□ Biceps brachii

□ Triceps brachii

□ Latissimus dorsi

□ Abdominal and hindlimb muscles

□ External oblique

□ Rectus abdominis

□ Tensor fasciae latae

□ Gluteus maximus

□ Biceps femoris

□ Semitendinosus

□ Gastrocnemius

□ Gracilis

□ **Cardiovascular Structures**

□ Heart

□ Left and right atria

□ Left and right ventricles

□ Aorta

☐ Major aortic branches

☐ Common carotid arteries

☐ Brachiocephalic artery

☐ Left subclavian artery

☐ Celiac artery (trunk)

☐ Superior mesenteric artery

☐ Renal arteries

☐ Gonadal arteries

☐ Common iliac arteries

☐ Anterior vena cava (human: superior vena cava)

☐ Subclavian veins

☐ Posterior vena cava (human: inferior vena cava)

☐ Hepatic vein

☐ Renal veins

☐ Common iliac veins

☐ Hepatic portal vein

☐ **The Viscera**

☐ Respiratory structures

☐ Larynx

☐ Trachea

☐ Primary bronchi

☐ Lungs

☐ Pleurae

☐ Diaphragm

☐ Digestive structures

☐ Submandibular salivary glands

☐ Esophagus

☐ Stomach

☐ Liver

☐ Pancreas

☐ Small intestine

☐ Mesentery

☐ Cecum

☐ Colon

☐ Lymphatic organs

☐ Spleen

☐ Thymus

☐ Urinary structures

☐ Kidney

☐ Ureter

☐ Urinary bladder

☐ Urethra

☐ Endocrine glands

☐ Thyroid gland

☐ Thymus gland

☐ Pancreas

☐ Adrenal glands

☐ Testes

☐ Ovaries

☐ Male reproductive organs

☐ Testes

☐ Epididymis

☐ Ductus (vas) deferens

☐ Seminal vesicles

☐ Prostate gland

☐ Penis

☐ Female reproductive organs

☐ Ovaries

☐ Uterine tubes (oviducts)

☐ Uterus

☐ Vagina

☐ Pregnant uterus

☐ Amniotic sac

☐ Embryo (or fetus)

☐ Umbilical cord

☐ Placenta

Notes or sketches:

LAB EXERCISE 42

The Fetal Pig

This exercise may be used in addition to, or in place of, the laboratory rat dissection presented in Lab Exercise 41. It can serve either as an introduction to vertebrate anatomy or as a concluding synthesis activity.

Hint . . . In a short course, a well-preserved specimen can be used from time to time throughout your studies. If you will be looking at your dissected specimen from time to time for the next several months, make sure it is kept in the appropriate container under conditions suggested by your instructor.

Before You Begin

☐ Set your learning goals. When you finish this exercise, you should be able to:
  o perform a whole-body dissection of a vertebrate animal
  o identify the major anatomical features of the vertebrate body in a dissected specimen

☐ Prepare your materials:
  o preserved (plain or double-injected) fetal pig
  o dissection tools and tray
  o mounted fetal pig skeleton (optional)
  o storage container (if specimen is to be reused)

☐ **Carefully** read the directions and safety tips for this exercise before starting any procedure.

SAFETY FIRST!

Observe the usual precautions when working with a preserved or fresh specimen. Heed the safety advice accompanying preservatives used with your specimen. Use protective gloves and safety eyewear while handling your specimen. Avoid injury with dissection tools. Dispose of your specimen as instructed.

A. The External Aspect

Examine the external aspect of your specimen:

☐ **1** Determine the anatomical orientation of the specimen. Which direction is anterior? Posterior? Which direction is ventral? Dorsal? Identify sagittal, transverse, and frontal planes in your specimen.

☐ **2** Identify these externally visible features:
  o **pinna (auricle)**
  o **external nares (nostrils)**
  o **integument**
  o **umbilicus (umbilical cord)**
  o **forelimbs**
  o **hindlimbs**
  o **thoracic region**
  o **abdominal region**
  o **nipples**
  o **anus**
  o **tail**

☐ **3** Determine the sex of your specimen by examining the external genitals:
  o **Female**—Immediately anterior to the anus, on the ventral surface, is the **vulva** with an opening to the **vagina** and the **urethra.** The vulva is also called the *urogenital opening.*
  o **Male**—In the male fetal pig, there is a rather loose area of skin immediately posterior to the anus, perhaps even hiding the anus from view. Around the time of birth, each **testis** will descend from its position inside the body into the space under this skin. The skin will pouch out to form the outer wall of the **scrotum.** Just posterior to the umbilical cord is the distal end of the **penis** with its **prepuce,** or skin-fold covering. Locate the opening of the **urethra** in the penis.

Hint . . . Figures 42.1–42.5 provide a complete set of anatomical sketches of the fetal pig.

B. Skin, Bones, and Muscles

□ **1** Remove most of the skin from your specimen by following this procedure:
- o Place the animal in the tray with its ventral surface facing you.
- o Pull up on the skin over the neck and puncture it with the tip of a scissors. Slide the bottom tip of the scissors into the **subcutaneous** area under the skin.
- o Begin cutting along the lines indicated in figure 42.1*a*. Notice that the posterior cuts are different for male and female specimens.
- o Be careful not to cut into the skeletal muscles under the skin or through the base of the umbilical cord.
- o With your forceps, pull the two flaps of skin over the neck away from the animal's body. Notice the **areolar tissue** under the skin that is pulled apart as you remove the skin. Sometimes it helps if you scrape at the loose connective tissue under the skin with your scalpel as you peel the skin away. The skinning process is difficult unless you have patience and proceed slowly. Pull the flaps of skin over the abdomen and over the **groin** area away in a similar fashion. If you have a male pig, leave the skin posterior to the umbilical cord in place for now.
- o Pin the flaps of skin to the floor of the dissection tray or cut them away from the body entirely. *If you plan to study the musculature of the dorsum you must similarly remove the skin from the animal's back.* (See Lab Exercise 41, figure 41.2, for an alternate cutting pattern.)

□ **2** Examine the skin, identifying these features:
- o **dermis**—the thick, inner layer of the skin
- o **epidermis**—the thinner, outer layer of the skin
- o **hypodermis**—the *subcutaneous tissue* under the skin proper, made of areolar and adipose tissue

□ **3** Explore the shape of the skinned fetal pig body. Which **bones** of the fetal pig's skeleton can you see or palpate (feel)? If you have a mounted fetal pig skeleton available, identify as many of the bones of the skeleton as you can. If you become stumped, refer to figure 42.1*b*. Notice the similarity between the fetal pig's skeletal plan and that of the human. Name some differences. One difference that is easy to see is in the vertebral column. Although the number of seven cervical vertebrae is the same for both, the other vertebrae differ in number. Note that there are numerous **caudal vertebrae** instead of a single

coccyx. Also, because the fetal pig's skeleton is just beginning its development, many of the bones are very cartilaginous rather than bony.

□ **4** Observe the fetal pig's musculature. Some of the external muscles of the torso can be separated from each other for easier viewing. Slide a probe into the loose connective tissue joining adjacent muscles and run the probe along their margins. Using figure 42.2 as a guide, try to identify the major skeletal muscles of the fetal pig's body. Because the animal was not yet active when the specimen was prepared, the muscles appear underdeveloped. This will make muscle identification difficult.

C. Cardiovascular Structures

□ **1** Open the **ventral body cavity** by cutting into its muscular wall in a manner similar to your earlier cut into the skin over the abdomen. Cut flaps in the neck, abdomen, and groin areas as you did with the skin. Be careful not to damage any **visceral organs** with your scissors as you cut. Likewise, avoid injuring the internal structures associated with the umbilicus. Fold back the flaps and anchor them with pins.

□ **2** Locate the **heart** near the middle of the **thoracic cavity,** in the **mediastinum.** Can you identify the four chambers? After identifying the major vessels (see steps 3 and 4), you may want to remove the heart and dissect it as you did with the sheep heart earlier in the course. See Lab Exercise 26 for more details.

□ **3** If you have a double-injected preserved specimen, the systemic **arteries** are filled with red latex and the systemic **veins** are filled with blue latex. If not, the arteries can usually be distinguished from veins because they are stiffer and lighter in color than veins. Locate the **aorta,** the large artery leaving the heart and arching posteriorly. Trace the major branches of the aorta, naming them if you can. Use figure 42.3 if you need help.

□ **4** Locate the **anterior vena cava** where it drains into the heart. It is sometimes called the *precava* and is analogous to the superior vena cava in the human. Follow its tributary veins, and identify them with the help of figure 42.3. Locate the **posterior vena cava** (*postcava*) and trace its tributaries.

Hint . . . Once you have cut into the ventral body cavity, you may be tempted to cut and remove organs. It is important that you keep everything as intact as possible. You may pull organs to one side or another to view deeper structures, but avoid making cuts.

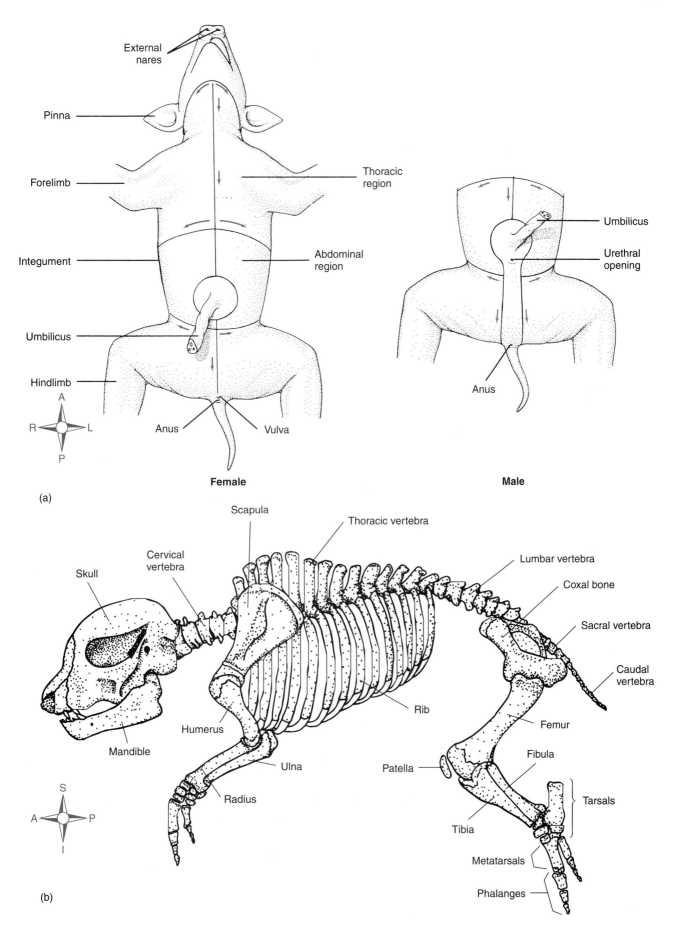

Figure 42.1　(*a*) External aspect of the female (left) and male (right). (*b*) Fetal pig skeleton.

42-4

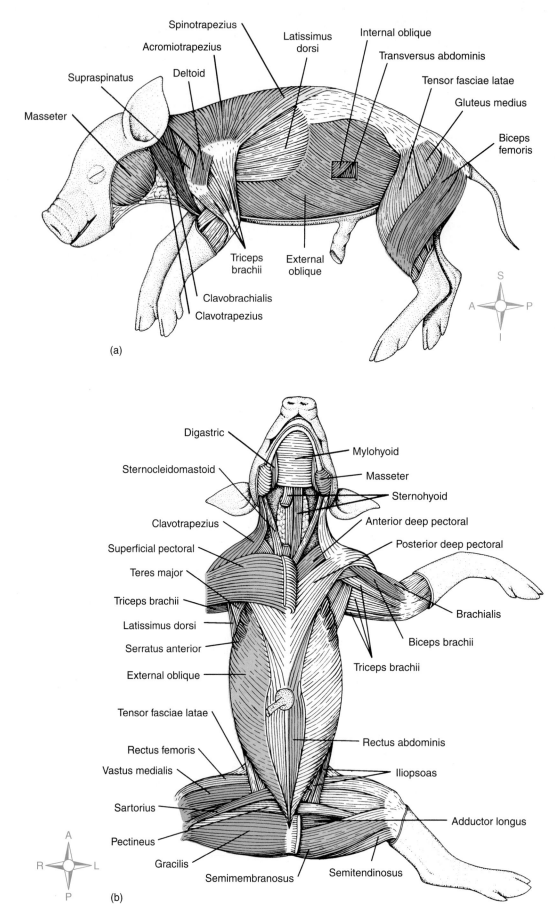

Figure 42.2 (*a*) Lateral view of fetal pig musculature. (*b*) Ventral view of fetal pig musculature.
316

(a)

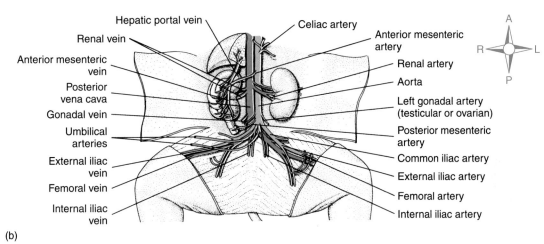

(b)

Figure 42.3 (*a*) Arteries (left) and veins (right) of the anterior body. (*b*) Major vessels of the posterior body.

D. The Viscera

The viscera, or major internal organs, can be seen within the ventral body cavity. Use figures 42.2, 42.3, and 42.4 to guide you in locating the following:

☐ **1**   Locate some of these features of the lower **respiratory system:**
- o **larynx**
- o **trachea**
- o **primary bronchi**
- o **lungs** (can you distinguish the parietal from visceral **pleurae?**)
- o **diaphragm**

☐ **2**   Locate these structures of the **digestive system:**
- o **esophagus**
- o **stomach**
- o **liver** (four separate lobes)
- o **gallbladder**
- o **pancreas**
- o **small intestine**
- o **mesentery**
- o **large intestine** (*spiral colon*)

☐ **3**   Locate these **lymphatic organs:**
- o **spleen**
- o **thymus**

☐ **4**   Locate these features of the **urinary system** (fig. 42.5):
- o **kidney**
- o **renal cortex**
- o **renal pyramid**
- o **renal pelvis**
- o **renal calyx**
- o **ureter**
- o **urinary bladder**
- o **urethra**

☐ **5**   Try to locate these **endocrine glands** in your specimen:
- o **thyroid gland**
- o **thymus gland**
- o **pancreas**
- o **adrenal glands**
- o **testes**
- o **ovaries**

☐ **6**   Identify these structures associated with the **male reproductive system:**
- o **testes**
- o **epididymis**
- o **ductus (vas) deferens**
- o **seminal vesicle**
- o **penis**
- o **scrotum**

☐ **7**   Find these **female reproductive system** structures (fig. 42.5):
- o **ovaries**
- o **oviducts** (fallopian tubes)
- o **uterus** (notice that the fetal pig uterus has a Y shape, with a right and left *uterine horn*)
- o **vagina**

Hint . . . Unless your lab group has both a male and a female specimen, you may want to temporarily trade specimens with a group that has a fetal pig of a different gender from yours. By doing so, you will be able to find the features of both reproductive systems.

☐ **8**   Carefully examine the umbilical cord, noting these structures:
- o **umbilical vein**
- o **umbilical arteries**

Sketch a cross section of the umbilical cord in the lab report and label all identifiable structures.

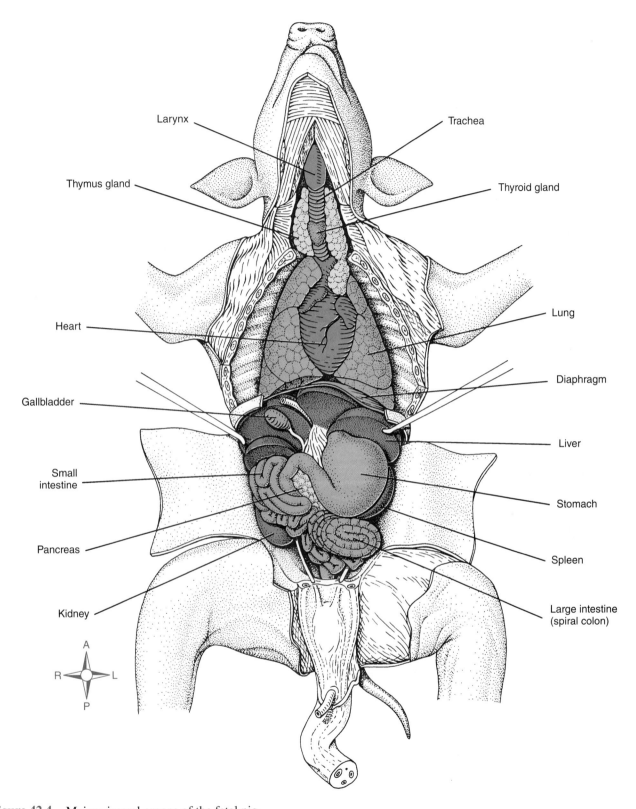

Figure 42.4 Major visceral organs of the fetal pig.

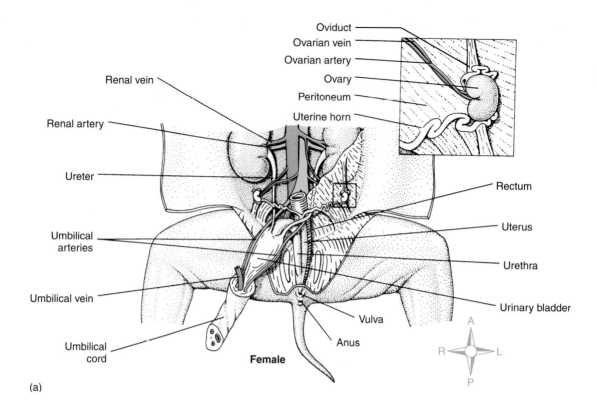

(a)

Oviduct
Ovarian vein
Ovarian artery
Ovary
Peritoneum
Uterine horn

Renal vein
Renal artery
Ureter
Umbilical arteries
Umbilical vein
Umbilical cord

Rectum
Uterus
Urethra
Urinary bladder
Vulva
Anus

Female

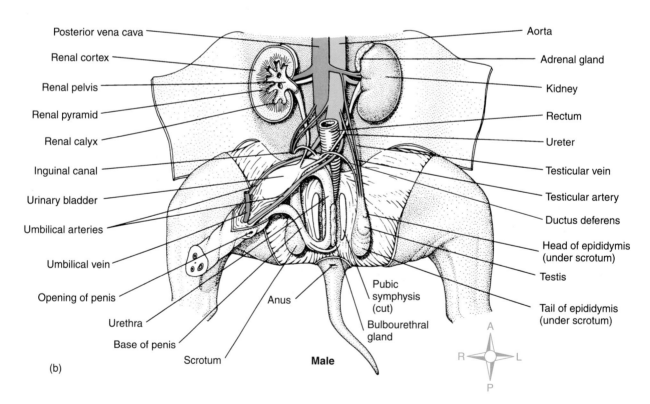

(b)

Posterior vena cava
Renal cortex
Renal pelvis
Renal pyramid
Renal calyx
Inguinal canal
Urinary bladder
Umbilical arteries
Umbilical vein
Opening of penis
Urethra
Base of penis
Scrotum
Anus

Aorta
Adrenal gland
Kidney
Rectum
Ureter
Testicular vein
Testicular artery
Ductus deferens
Head of epididymis (under scrotum)
Testis
Tail of epididymis (under scrotum)

Pubic symphysis (cut)
Bulbourethral gland

Male

Figure 42.5   Urinary and reproductive organs of (*a*) the female and (*b*) the male.

LAB REPORT 42

The Fetal Pig

Name _____

Date _____ Section _____

Fetal Pig Dissection Checklist

☐**External Aspect**

☐Pinna (auricle)

☐External naris (nostril)

☐Integument

☐Umbilicus (umbilical cord)

☐Forelimbs

☐Hindlimbs

☐Thoracic region

☐Abdominal region

☐Nipples

☐Anus

☐Tail

Gender: ☐ female

☐ Vulva (urogenital opening)

Gender: ☐ male

☐ Opening of penis

☐**Skin, Bones, and Muscles**

☐Dermis

☐Epidermis

☐Hypodermis

☐Axial skeleton

☐ Cranial bones

☐ Facial bones

☐ Cervical vertebrae

☐ Thoracic vertebrae

☐ Lumbar vertebrae

☐ Sacral vertebrae

☐ Caudal vertebrae

☐ Ribs

☐ Sternum

☐Appendicular skeleton

☐ Scapula

☐ Clavicle

☐ Humerus

☐ Radius

☐ Ulna

☐ Carpal bones

☐ Metacarpal bones

☐ Phalanges of the forelimb

☐ Coxal bone

☐ Femur

☐ Patella

☐ Tibia

☐ Fibula

☐ Tarsal bones

☐ Metatarsal bones

☐ Phalanges of the hindlimb

☐Head, neck, and shoulder muscles

☐ Masseter

☐ Sternomastoid (human: sternocleidomastoid)

☐ Trapezius

☐ Deltoid

☐ Biceps brachii

☐ Triceps brachii

☐ Latissimus dorsi

☐ Brachialis

☐ Superficial pectoral (human: pectoralis major)

☐ Anterior/posterior deep pectorals (human: pectoralis minor)

☐Abdominal and hindlimb muscles

☐ Teres major

☐ Serratus anterior

☐ External oblique

☐ Internal oblique

☐ Transversus abdominis

☐ Rectus abdominis

☐ Tensor fasciae latae

☐ Iliopsoas

☐ Rectus femoris

☐ Vastus medialis

☐ Gluteus

☐ Sartorius

☐ Biceps femoris

☐ Semimembranosus

☐ Semitendinosus

☐ Gastrocnemius

☐ Adductor longus

☐ Gracilis

☐**Cardiovascular Structures**

☐Heart

☐ Left and right atria

☐ Left and right ventricles

☐ Coronary vessels

☐Aorta

☐Major aortic branches

☐ Common carotid arteries

☐ Brachiocephalic arterial trunk

☐ Left subclavian artery

☐ Celiac artery (trunk)

☐ Superior mesenteric artery

☐ Renal arteries

☐ Gonadal arteries

☐ Common iliac arteries

☐ Anterior vena cava (precava)
(human: superior vena cava)

☐ Internal jugular veins

☐ External jugular veins

☐ Subclavian veins

☐ Posterior vena cava (postcava)
(human: inferior vena cava)

☐ Renal veins

☐ Gonadal veins

☐ Common iliac veins

☐ Internal iliac veins

☐ External iliac veins

☐ Femoral veins

☐ Hepatic portal vein

☐ **The Viscera**

☐ Respiratory structures

☐ Larynx

☐ Trachea

☐ Primary bronchi

☐ Lungs

☐ Pleurae

☐ Diaphragm

☐ Digestive structures

☐ Esophagus

☐ Stomach

☐ Liver

☐ Gallbladder

☐ Pancreas

☐ Small intestine

☐ Mesentery

☐ Large intestine (spiral colon)

☐ Lymphatic organs

☐ Spleen

☐ Thymus

☐ Urinary structures

☐ Kidney

☐ Renal cortex

☐ Renal pyramid

☐ Renal pelvis

☐ Renal calyx

☐ Ureter

☐ Urinary bladder

☐ Urethra

☐ Endocrine glands

☐ Thyroid gland

☐ Thymus gland

☐ Pancreas

☐ Adrenal glands

☐ Testes

☐ Ovaries

☐ Male reproductive organs

☐ Testes

☐ Epididymis

☐ Ductus (vas) deferens

☐ Seminal vesicles

☐ Penis

☐ Scrotum

☐ Female reproductive organs

☐ Ovaries

☐ Oviducts (fallopian tubes)

☐ Uterus, including uterine horns

☐ Vagina

☐ Vulva

☐ Umbilical cord

☐ Umbilical vein

☐ Umbilical arteries

Umbilical cord sketch:

LAB EXERCISE 43

The Human

This exercise is intended as a possible summary or synthesis activity for the short laboratory course. Whether or not your instructor will invite you to complete this exercise will depend on the course schedule and the availability of the required materials.

The first activity of this exercise suggests that you observe a live or videotaped demonstration of human anatomy using a human body. The second activity presents a special focus on transverse (horizontal) sectional anatomy, which is becoming increasingly important in practical applications of human anatomy.

Before You Begin

☐ Read the appropriate chapter in your textbook.

☐ Set your learning goals. When you finish this exercise, you should be able to:
  o describe the overall body plan of the human
  o identify the major organs of the human in a previously dissected cadaver or in a human body model or chart
  o identify structures in selected transverse sections of the human body

☐ Prepare your materials:
  o prosected human cadaver (for demonstration, if available) or human dissection videotaped film
  o plastinated preparation (or chart): *human thorax, horizontal section (just above heart)*
  o plastinated preparation (or chart): *human abdomen, horizontal section (pancreas level)*
  o demonstration pointers
  o hand lens
  o ANATOMY AND PHYSIOLOGY REVEALED 2.0 and computer system (optional)

☐ **Carefully** read the directions and safety tips for this exercise before starting any procedure.

A. The Human Cadaver

Advanced laboratory courses in human anatomy use the human cadaver as the basic dissection specimen. A **cadaver** is an embalmed corpse. If available, view a live or videotaped demonstration using a prosected human cadaver. A prosected specimen is one that has been previously dissected and prepared for anatomical demonstrations.

SAFETY FIRST!

Do not touch the cadaver unless invited to do so by the demonstrator. If you are invited to touch it, use gloved hands. Cadavers are normally fixed in preservatives that are toxic. Do not inhale the fumes that are immediately above the specimen. Excuse yourself from the demonstration if you feel ill or if you are sensitive to preservatives, such as formalin. Pregnant women should avoid contact with formalin fumes.

B. Transverse Sectional Anatomy

Clinical and research institutions are seeing an increased use of computed tomography (CT) scans, magnetic resonance (MR) imaging, ultrasonography, and other advanced imaging techniques. Each of these techniques requires a basic knowledge of human anatomy as seen in a transverse, or horizontal, section.

Your studies so far have emphasized the sagittal and frontal aspects of anatomy. Now that you are familiar with the essentials of human anatomy from those perspectives, it is time to apply that knowledge to locating structures in transverse preparations.

SAFETY FIRST!

Plastinated specimens are natural tissues that have been embedded in plastic. Despite the fact that they long outlast normally embalmed specimens, they are very fragile. If you handle them at all, be careful not to damage them. Always use a demonstration pointer in pointing to structures, never a pen or pencil. Do not touch the specimen with your pointer; just point at it.

☐ **1** Obtain a plastinated transverse section of a human torso just superior to the heart (at about T4). Try to locate these structures (fig. 43.1):

- **Aorta**—If the aorta is present, what section is it (ascending, arch, or descending)?
- **Superior vena cava**—Can you identify this vein by its position and the thinness of its wall?
- **Esophagus**—This is a small (and perhaps flattened) muscular tube anterior to a vertebra.
- **Trachea**—Are any of the C-shaped cartilage rings visible in the tracheal wall?
- **Thymus**—It is posterior to the sternum.
- **Lungs**—Can you distinguish any portions of the respiratory tract within the lung tissue? Identify the parietal and visceral pleurae and the pleural cavity.
- **Sternum**—This structure is in the anterior wall of the thoracic cavity.
- **Ribs**
- **Thoracic vertebra**—What features of the vertebra are visible in your specimen?
- **Spinal cord**—It is inside the vertebral foramen. What features of the spinal cord can you identify?
- **Skeletal muscles**—They form the wall of the thoracic cavity.
- **Subcutaneous tissue**—Identify the areolar and/or adipose tissue under the skin.
- **Integument**—Can you distinguish between the dermis and the epidermis?

Are there any other organs or structures visible in your specimen? If the upper arms are present in your specimen, identify the humerus and other arm structures.

> *Hint . . .* Use a hand lens to examine the detail of the structures you have found.

☐ **2** Obtain a plastinated transverse section of the human abdomen at the level of the pancreas (about T12). Try to locate these structures (fig. 43.2):

- **Vertebra**—What features are visible in this preparation?
- **Spinal cord**—Can you distinguish white from gray matter?
- **Kidneys**—Are they the same size? Why or why not? Identify the renal fat pad around each kidney. Can you distinguish between the renal cortex and the renal medulla?
- **Aorta**—The abdominal aorta may be present just anterior to the body of the vertebra.
- **Inferior vena cava**—This large vein is normally next to the aorta.
- **Pancreas**—This glandular tissue is near the stomach.
- **Stomach**—This large, muscular organ should be present in the left portion of the specimen. Can you identify the rugae (large folds) of the gastric mucosa?
- **Liver**—This is the dominant feature of your specimen, a large mass of tissue to the right and anterior of the abdominal cavity.
- **Colon**—Is any portion of the colon visible in your specimen? If so, which section is it?
- **Peritoneum**—Can you distinguish between the parietal and visceral portions of the peritoneum? Identify the peritoneal cavity. Are any mesenteries distinguishable in your specimen? The lesser omentum?
- **Ribs**
- **Abdominal muscles**—Which abdominal muscles could be present in your section?
- **Subcutaneous tissue**—Is the adipose tissue in this layer of the same thickness all the way around the abdominal wall?
- **Integument**—Are the dermis and epidermis distinguishable in this section?

> *Hint . . .* Each specimen is unique, just as every person's body is unique. Therefore, some structures listed may not be present (or identifiable) in your specimen.

Transverse Thoracic Section

AORTA 1
SUPERIOR VENA CAVA 2
ESOPHAGUS 3
TRACHEA 4
THYMUS 5
LUNG 6
PLEURAL CAVITY 7

STERNUM 8
RIB 9
VERTEBRA 10
SPINAL CORD 11
SKELETAL MUSCLE 12
SUBCUTANEOUS TISSUE 13
INTEGUMENT 14

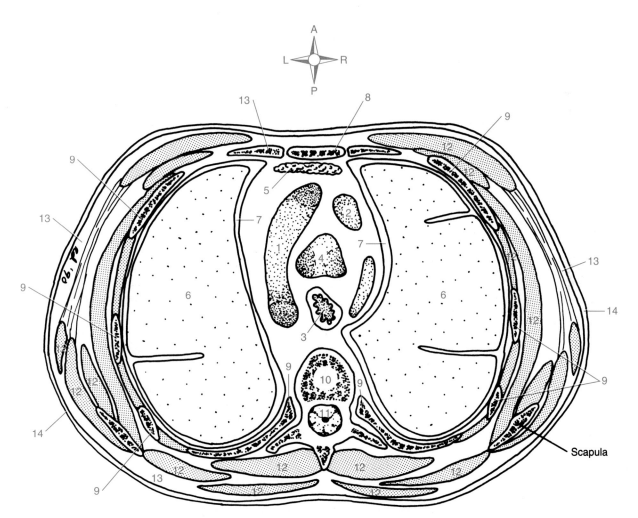

Figure 43.1   Color the major features of this transverse section of the human thorax.

Transverse Abdominal Section

VERTEBRA 1
SPINAL CORD 2
KIDNEY 3
AORTA 4
INFERIOR VENA CAVA 5
PANCREAS 6
STOMACH 7

LIVER 8
COLON 9
PERITONEAL CAVITY 10
RIB 11
ABDOMINAL MUSCLE 12
SUBCUTANEOUS TISSUE 13
INTEGUMENT 14

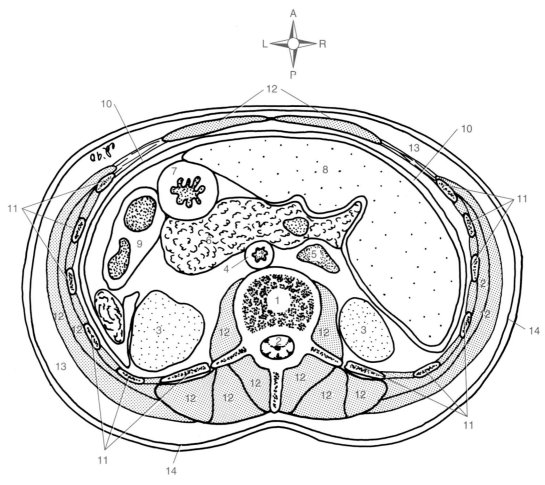

Figure 43.2   Color the major features of this transverse section of the human abdomen.

The Human

Name _____

Date _____ Section _____

Sketch and label your thorax (transverse section) specimen:

Sketch and label your abdomen (transverse section) specimen:

NOTES

NOTES

NOTES

NOTES

NOTES

NOTES

NOTES

NOTES

NOTES

NOTES